T0259008

Mathematical Foundations of Public Key Cryptography

Mathematical Foundations of Public Key Cryptography

Xiaoyun Wang
Guangwu Xu
Mingqiang Wang
Xianmeng Meng

CRC Press
Taylor & Francis Group
Boca Raton London New York

CRC Press is an imprint of the
Taylor & Francis Group, an **informa** business

CRC Press
Taylor & Francis Group
6000 Broken Sound Parkway NW, Suite 300
Boca Raton, FL 33487-2742

© 2016 by Science Press
CRC Press is an imprint of Taylor & Francis Group, an Informa business

No claim to original U.S. Government works

Printed on acid-free paper
Version Date: 20150813

International Standard Book Number-13: 978-1-4987-0223-2 (Hardback)

Visit the Taylor & Francis Web site at
http://www.taylorandfrancis.com

and the CRC Press Web site at
http://www.crcpress.com

Contents

Foreword

As a cornerstone of information security, cryptography is a subject with an ancient history, but it also is an emerging discipline that is widely and effectively used in many areas of modern society. One of the two fundamental issues of cryptography is to securely encrypt information so that a third party will not get the content from its encrypted form. The other one, in contrast, is how to break the encryption and get the information from its encrypted form. There are many methods that can be used in cryptography to protect information and break codes. Mathematics has always been an important tool in cryptography. With the invention of public key cryptosystem, mathematics has been playing an indisputably important role in cryptography, so cryptography has also become a special subject of mathematics.

Professor Xiaoyun Wang has always attached great importance to training students in information security and building its core curricular structure. "Number Theory and Algebraic Structures" is a core course for cryptography, and she started to write lecture notes for the course as early as 2003. These lectures were successfully taught at both Shandong University and Tsinghua University, China. In collaboration with Guangwu Xu, Minqiang Wang, and Xianmeng Meng, Professor Wang revised and extended these lecture notes and published them in this book. Since the content of this book has been described in detail by the authors in the preface, I am not going to go over that again here. However, I do want to point out that this book has a distinctive feature; namely, it closely integrates basic number theory and algebra into cryptographic algorithms and complexity theory in a well-organized manner throughout this book. This is of great importance for training students in cryptography to have a unique way of thinking. My feeling is that the "formalization," which has been a quite effective way of thinking in mathematics, is hard to work directly in cryptography as it seems that the "formalization" thinking does not lead one to the essence of cryptographic problems. This is a key problem that needs special attention for mathematicians who decide to shift their research to cryptography. Therefore, this book will definitely play a very positive role in improving the quality of core courses in information security curriculum.

Professor Xiaoyun Wang studied with my brother, Professor Chengdong Pang, a strong advocate in applying mathematics to science and technology. He participated in the preliminary research of seepage theory, thin shell theory, spline theory and application, directional blasting, and so on. Around 1990, he spearheaded the effort to establish a cryptography research group and train graduate students at Shandong University, China. Professor Xiaoyun Wang was the most accomplished researcher in this rather successful group. In the field of cryptanalysis, she has proposed and developed a theory and technique for collision attack of hash functions, successfully breaking some major popular cryptographic hash functions, such as MD5 designed by the Turing award recipient Rivest and SHA-1 designed by the National Institute of Standards and Technology (NIST) and the National Security Agency (NSA) of the United States. These hash functions were the core components of commonly used digital signatures and digital certificates. Professor Wang's research has surprised the cryptography community, prompting the NIST to initiate a 5-year project for the new standard hash function SHA-3 in 2007. She has also made great contributions in analyzing message authentication codes and designing cryptographic algorithms. She led the design of hash algorithm SM3, which has been adopted as the national commercial hash algorithm.

Although I am not an expert in cryptography, I am delighted to write the Foreword to Xiaoyun Wang's book, as I feel it is my responsibility and obligation to do so. I hope that she will continue to passionately dedicate herself to research and teaching without being distracted by fame and gain.

Chengbiao Pan

Preface

Ever since Diffie and Hellman proposed the idea of public key cryptography, cryptographers have designed many topical public key algorithms. The security of all of these algorithms is based on some classical hard problems in mathematics, for examples, the integer factorization problem, the discrete logarithm problem, the knapsack problem, and the shortest vector problem in a lattice. The study of fast algorithms to solve these hard mathematical problems lies at the heart of the cryptanalysis of public key algorithms. In order for students in information security to successfully and effectively grasp the basic theory of modern cryptography and have a deeper understanding of the interdisciplinary nature of cryptography and mathematics, we wrote *Mathematical Foundations of Public Key Cryptography* as a textbook to help students lay a solid foundation in mathematics for their future study. The theoretical knowledge involved in this book includes the fundamentals of mathematics necessary for modern cryptography, especially public key cryptography. Therefore, this book serves as a textbook for undergraduate students in information security and also as a reference book for professionals in cryptography.

The lecture notes "Number Theory and Algebraic Structures," which the authors started to write when the information security program at Shandong University, China was founded in 2003, contain the basic mathematical knowledge needed for modern cryptography, especially public key cryptography. Rather than simply combining number theory and modern algebra, this book features the interdisciplinary characteristics of cryptography and reveals the integrations of mathematical theories and cryptographic applications. This book has three distinguishing features: First, the basic content of number theory and algebra covers the important mathematical concepts necessary for cryptography. For instance, we have introduced some fundamental concepts and methods in elementary number theory, such as the division algorithm, Euler's theorem, the Chinese remainder theorem, and primitive roots; we have also described some widely used mathematical theories and methods for cryptography, such as finding greatest common divisors using the division algorithm, the operation of computing modulo inverse, discrete logarithms, and integer factorization. Second, we emphasize the close integration of theory and practice while paying attention to the practical side. We provide a sufficient number of practical exercises when we discuss important algorithms so that students will understand the applications of theory in practical situations. The third feature is the incorporation of the complexity theory of algorithms throughout this book by introducing the basic number theoretic and algebraic algorithms and their complexities, so that readers would have some preliminary understanding of the applications of mathematical theories in cryptographic algorithms.

This book consists of 11 chapters. Basic theory and tools of elementary number theory, such as congruences, primitive roots, residue classes, and continued fractions, are covered in Chapters 1 through 6. Knowledge of primitive roots is the theoretical background required for the Diffie–Hellman public key algorithms, while continued fractions have important applications in the analysis of RSA public key algorithms as well as in integer factorization. The basic concepts of abstract algebra are introduced in Chapters 7 through 9, where three basic algebraic structures of groups, rings, and fields and their properties are explained. The Chinese remainder theorem, which has significant applications in the big integer multiplications and efficient implementation of cryptographic algorithms, is also covered in detail. Chapter 10 is about the basic theory of complexity and several related mathematical algorithms, including primality testing, the discrete logarithm problem, and the integer factorization problem. Chapter 11 presents the basics of lattice theory and the lattice basis reduction algorithm—the LLL algorithm and its application in the cryptanalysis of the RSA algorithm.

The early version of this book was used many times as lecture notes for information security majors at Shandong University. Based on the feedback, the authors revised and updated the lecture notes, which were eventually developed into the current version of this book. In Chapters 1 to 6, we mainly referenced *Elementary Number Theory* by Professors Chengdong Pan and Chengbiao Pan [1]. In Chapters 7 to 9, we referenced three independently authored textbooks on modern algebra, by Professors Pinsan Wu [2], Herui Zhang [3], and Shaoxue Liu [4] respectively.

Although we have strived to do our best with this book, it is inevitable that this book might still contain places of imperfections, so any suggestions or feedback will be greatly appreciated.

Xiaoyun Wang
Guangwu Xu
Mingqiang Wang
Xianmeng Meng

Acknowledgments

The publication of this book was made possible by the China Education Ministry's project on supporting distinguished curriculum in information security, the key grant of the National Science Foundation (No. 61133013), and the National Basic Research Program of China (973 Program; No. 2013CB834205). Professor Tao Zhan has given us much advices and suggested the current title of this book. Shihui Zheng, Qi Feng, and Puwen Wei have provided valuable inputs in the process of writing this book. We would like to express our gratitude to everyone who has helped, supported, and inspired us from the inception to the publication of this book.

CHAPTER 1

Divisibility of Integers

D IVISIBILITY IS a key concept in number theory. The main purpose of this chapter is to introduce some basic concepts and properties that relate to divisibility. Some of them, such as divisibility, divisors, common divisors, the least common multiples, and factorization, have already been taught in middle school and high school. Here we shall define them by a precise mathematical language. Through mastering mathematical definitions and properties of these concepts, we can further solve many elementary number theoretic problems related to divisibility. The theory of divisibility has a rich content and provides flexible problem-solving methods. It not only is the foundation of number theory and algebra, but also has a wider range of applications to cryptography. Some important applications in cryptography include prime factorization of integers, and the Euclidean algorithm for finding greatest common divisors.

1.1 THE CONCEPT OF DIVISIBILITY

We use $\mathbb{Z}$ to denote the set of all integers and $\mathbb{N}$ the set of all natural numbers. The definition of divisibility is as follows:

Definition 1.1 Let $a, b \in \mathbb{Z}$. If there is a $q \in \mathbb{Z}$, such that $b = aq$, then b is said to be divisible by a and denoted as $a|b$; b is called a multiple of a, and a is called a divisor (or a factor) of b. Otherwise, b is said to be not divisible by a, or a does not divide b, denoted by $a \nmid b$.   □

By this definition and the law of operation, the following properties of divisibility are immediate:

Theorem 1.2 *Let $a, b, c \in \mathbb{Z}$. Then*

1. *$a|b$ and $b|c \Rightarrow a|c$.*

2. *$a|b$ and $a|c \Leftrightarrow$ for all $x, y \in \mathbb{Z}$, $a|bx + cy$.*

3. *If $m \in \mathbb{Z}$ and $m \neq 0$, then $a|b \Leftrightarrow ma|mb$.*

4. *$a|b$ and $b|a \Rightarrow a = \pm b$.*

5. *If $b \neq 0$, then $a|b \Rightarrow |a| \leq |b|$.*

Proof

1. Since $a|b$, by the definition of divisibility, there is a q_1 such that $b = aq_1$. Similarly, there is a q_2 such that $c = bq_2$. Therefore,

$$c = q_2 b = (q_1 q_2)a,$$

   or $a|c$.

2. From $a|b, a|c$, we see that there exist r, s such that $b = ar, c = as$. For any $x, y \in \mathbb{Z}$, we have

$$bx + cy = arx + asy = a(rx + sy),$$

   that is, $a|bx + cy$.

   Conversely, assume that for all $x, y \in \mathbb{Z}$, $a|bx + cy$. We get $a|b$ and $a|c$ immediately by choosing $x = 1, y = 0$ and $x = 0, y = 1$, respectively.

   The proofs for properties 3, 4, and 5 are similar and left to the reader.

   Obviously, $\pm 1, \pm b$ are factors of b, and we call them the *trivial factors of b*; other factors of b are called *nontrivial factors* or *proper factors*. These lead us to the definition of prime.

Definition 1.3 Assume that $p \neq 0, \pm 1$. If p has no factors other than $\pm 1, \pm p$, then p is said to be a prime element (usually called a prime). An integer c is called a composite if c has a proper factor. □

Remark In general, primes in our consideration are restricted to be positive.

Next, we introduce some theorems concerning prime numbers.

Theorem 1.4 *If c is a composite, the least proper factor of c must be prime.*

Proof We know that $c > 2$ because c is a composite. Let d be a least proper factor of c. Suppose that d is not a prime; then there is a positive integer d' that is a proper factor d. From property 1, we see that $d' | c$. This contradicts the assumption that d is a least proper factor. The theorem is proved.

Theorem 1.5 *There are infinite many primes.*

Proof Assume that there are only finite many primes and they are $p_1, p_2, \cdots, p_k$. Consider $a = p_1 p_2 \cdots p_k + 1$. From Theorem 1.4, the least proper factor of a, denoted by p, must be a prime. Since p is prime, p has to be some p_i. This implies that $p | a$ and $p | p_1 p_2 \cdots p_k$; hence, $p | 1$. This contradicts the fact that p is a prime. Therefore, the theorem is proved.

Let us list all primes in ascending order and denote p_n the nth prime. Let $\pi(x)$ be the number of primes that are $\leq x$. Although we are not able to tell the exact location of p_n, we can get a weaker upper bound of it. The next theorem only describes a weaker upper bound of p_n and a weaker lower bound of $\pi(x)$. Finer estimations of $\pi(x)$ can be found in Chapter 5 of this book. Interested readers are also referred to [5].

Theorem 1.6 *We have the following results for the nth prime p_n and $\pi(x)$:*

1. $p_n \leq 2^{2^{n-1}}, n = 1, 2, \cdots$.

2. $\pi(x) > \log_2 \log_2 x, x \geq 2$.

Proof
1. We will use induction to prove the first part. The argument is obviously true when $n = 1$. Assume that the result holds for $n \leq k$. Then, for $n = k + 1$, by Theorem 1.4, $p_{k+1} \leq p_1 p_2 \cdots p_k + 1$. The induction hypothesis implies that

$$p_{k+1} \leq 2^{2^0} 2^{2^1} \cdots 2^{2^{k-1}} + 1 = 2^{2^k - 1} + 1 < 2^{2^k}.$$

Part 1 is proved.

2. For any $x \geq 2$, there must be a unique n such that $2^{2^{n-1}} \leq x < 2^{2^n}$. From part 1, we get

$$\pi(x) \geq \pi(2^{2^{n-1}}) \geq n > \log_2 \log_2 x.$$

The theorem is proved.

Another basic result in elementary number theory is the division algorithm. It is a generalization of divisibility and a foundation of the Euclidean algorithm.

Theorem 1.7 *Let a, b be two given integers with $a \neq 0$. Then there is a unique pair of integers q and r, such that*

$$b = qa + r, \quad 0 \leq r < |a|,$$

where r is called the least nonnegative remainder of b divided by a. Furthermore, $a|b$ if and only if $r = 0$.

Proof Uniqueness: If there are integers q' and r' with $0 \leq r' < |a|$ such that

$$b = aq' + r',$$

then we have $0 \leq |r' - r| < |a|$, and $r' - r = (q - q')a$. It is immediately evident that $r' - r = 0$ by the properties of divisibility, and the uniqueness follows.

Existence: If $a|b$, then we can set $q = \frac{b}{a}$, $r = 0$. For the case of $a \nmid b$, consider the set

$$T = \{b - ka : k = 0, \pm1, \pm2, \pm3, \cdots\}.$$

It is easy to see that T contains a positive integer (e.g., we can take $k = \frac{-2|ab|}{a}$); therefore, T must contain a smallest positive integer, and we denote such a number by

$$t_0 = b - k_0 a > 0.$$

We will prove that $t_0 < |a|$. Since $a \nmid b$, so $t_0 \neq |a|$. If $t_0 > |a|$, then $t_1 = t_0 - |a| > 0$. However, it is easy to see that $t_1 \in T$, $t_1 < t_0$, and this contradicts the minimality of t_0. To get the desired relation, we just take $q = k_0$, $r = t_0$.

It is obvious that $a|b$ if and only if $r = 0$.

The proof is complete.

We now partition all integers using the division algorithm. Let $a \geq 2$ be a given positive integer. For $j \in \{0, 1, \cdots, a-1\}$, the numbers that have remainder j, after dividing by a, are

$$ak + j, \quad k = 0, \pm 1, \pm 2, \pm 3, \cdots.$$

The set of these integers is denoted by $S_{a,j}$. The set $\{S_{a,j} : 0 \leq j \leq a-1\}$ satisfies the following two properties:

1. Any two distinct elements of $\{S_{a,j} : 0 \leq j \leq a-1\}$ are disjoint, that is,
$$S_{a,j} \cap S_{a,j'} = \emptyset, \ 0 \leq j \neq j' \leq a-1.$$

2. The union of all elements in $\{S_{a,j} : 0 \leq j \leq a-1\}$ is $\mathbb{Z}$, that is,
$$\underset{0 \leq j \leq a-1}{\cup} S_{a,j} = \mathbb{Z}.$$

Thus, integers are partitioned into a classes according to the least positive remainder after dividing by the number a. This kind of classification adds greater convenience in dealing with some mathematics problems.

Example 1.8 *For any integer x, the least nonnegative remainder of x^3 divided by 9 is one of $0, 1$, and 8.*

Proof For any integer x, there exists $0 \leq j \leq 8$ such that $x \in S_{9,j}$. So we only need to examine integers between 0 and 8:

$$\begin{array}{lll}
0^3 = 0 \times 9 + 0; & 1^3 = 0 \times 9 + 1; & 2^3 = 0 \times 9 + 8; \\
3^3 = 0 \times 9 + 0; & 4^3 = 7 \times 9 + 1; & 5^3 = 13 \times 9 + 8; \\
6^3 = 24 \times 9 + 0; & 7^3 = 38 \times 9 + 1; & 8^3 = 56 \times 9 + 8.
\end{array}$$

We get the desired result.

Example 1.9 *Let $a \geq 2$ be a given integer. Then any positive integer n can be uniquely written as*

$$n = r_k a^k + r_{k-1} a^{k-1} + \cdots + r_1 a_1 + r_0,$$

where the integers $k \geq 0$, $0 \leq r_j \leq a-1$ $(0 \leq j \leq k)$, $r_k \neq 0$. This is the radix-a representation of n. Such a representation is often written in the following form:

$$n = (r_k r_{k-1} \cdots r_1 r_0)_a.$$

Proof For the positive integer n there must be a unique $k \geq 0$ such that $a^k \leq n < a^{k+1}$. From the division algorithm, there is a unique pair q_0, r_0 such that

$$n = q_0 a + r_0, \ 0 \leq r_0 < a.$$

If $k = 0$, then $q_0 = 0, 1 \leq r_0 < a$; so the result holds. Assume that the result is true when $k = m \geq 0$. When it comes to the case of $k = m + 1$, the number q_0 from the previous relation must satisfy

$$a^m \leq q_0 < a^{m+1}.$$

By the hypothesis, this implies that

$$q_0 = s_m a^m + \cdots + s_0,$$

where $0 \leq s_j \leq a - 1 (0 \leq j \leq m - 1), 1 \leq s_m \leq a - 1$. Therefore,

$$n = s_m a^{m+1} + \cdots + s_0 a + r_0.$$

This means that the result is true for $m + 1$.

We remark that radix-10 the representation is just our usual integer's decimal representation. Another useful representation of integers is the radix-2 representation, which is also called the binary form.

Example 1.10 *Write the decimal integer* 27182 *as a radix-12 form.*

Solution: We perform the following calculations:

$$27182 = 12 \cdot 2265 + 2$$
$$2265 = 12 \cdot 188 + 9$$
$$188 = 12 \cdot 15 + 8$$
$$15 = 12 \cdot 1 + 3$$
$$1 = 12 \cdot 0 + 1$$

and get

$$27182_{10} = 13892_{12}.$$

1.2 THE GREATEST COMMON DIVISOR AND THE LEAST COMMON MULTIPLE

The greatest common divisor and the least common multiple are fundamental concepts in the theory of divisibility. This section defines the greatest common divisor and the least common multiple and discusses their properties.

Definition 1.11 Let a_1, a_2 be two integers. If $d|a_1$ and $d|a_2$, then d is said to be a common divisor (or common factor) of a_1 and a_2. More generally, let $a_1, a_2, \cdots, a_k$ be k integers. If $d|a_1, d|a_2, \cdots, d|a_k$, then d is said to be a common divisor of $a_1, a_2, \cdots, a_k$. □

Definition 1.12 Let a_1, a_2 be two integers (not both zero). Assume that d is a positive common divisor of a_1 and a_2. If, for any $d'|a_1, d'|a_2$, the relation $d'|d$ holds, then d is called the greatest common divisor of a_1 and a_2, and is denoted by $d = (a_1, a_2)$ or $d = gcd(a_1, a_2)$. More generally, let $a_1, a_2, \cdots, a_k$ be k integers (not all zero). Assume that d is a positive common divisor of $a_1, a_2, \cdots, a_k$, if, for any common divisor d' of $a_1, a_2, \cdots, a_k$, the relation $d'|d$ holds, then d is called the greatest common divisor of $a_1, a_2, \cdots, a_k$, and is denoted $(a_1, a_2, \cdots, a_k)$ or $gcd(a_1, a_2, \cdots, a_k)$. □

By Definition 1.12, the greatest common divisor is the largest one in the set of common divisors. See [1] for a proof of the existence of the greatest common divisor. The greatest common divisor has the following properties.

Theorem 1.13

1. *For any integer x, $(a_1, a_2) = (a_1, a_2 + a_1 x)$.*

2. *If $m > 0$, then $m(b_1, \cdots, b_k) = (mb_1, \cdots, mb_k)$.*

3. *If a_1, a_2 are two integers (not both zero), then*

$$\left(\frac{a_1}{(a_1, a_2)}, \frac{a_2}{(a_1, a_2)} \right) = 1;$$

more generally, if $a_1, \cdots, a_k$ are not all zero, then

$$\left(\frac{a_1}{(a_1, \cdots, a_k)}, \cdots, \frac{a_k}{(a_1, \cdots, a_k)} \right) = 1.$$

The arguments in the theorem are quite obvious, and their proofs are left to the reader. This theorem provides a simple and practical method for computing greatest common divisors.

Example 1.14 *For any integer m, we have*

$$(18m + 5, 12m + 3) = (6m + 2, 12m + 3) = (6m + 2, -1) = 1;$$

$$(4m + 5, 2m + 1) = (3, 2m + 1) = \begin{cases} 3, & m = 3k + 1, \\ 1, & m = 3k \ or \ 3k + 2. \end{cases}$$

Definition 1.15 Let a_1, a_2 be two integers. If $(a_1, a_2) = 1$, then a_1 and a_2 are said to be coprime. If $(a_1, \cdots, a_k) = 1$, then $a_1, \cdots, a_k$ are said to be coprime. □

Example 1.16 *Define the Fermat number as follows:*

$$F_n = 2^{2^n} + 1,$$

where n is a positive integer. Prove that any two distinct Fermat numbers are coprime.

Proof Given two distinct Fermat numbers F_m, F_n, assume $n = m + k, k > 0$. We have

$$\frac{F_n - 2}{F_m} = \frac{F_{m+k} - 2}{F_m} = \frac{2^{2^{m+k}} - 1}{2^{2^m} + 1} = \frac{(2^{2^m})^{2^k} - 1}{2^{2^m} + 1},$$

and hence, $F_m | F_n - 2$. This gives

$$(F_m, F_n) | F_n, (F_m, F_n) | F_n - 2,$$

so $(F_m, F_n) | 2$. By the definition of a Fermat number, F_n is an odd number; therefore, $(F_m, F_n) = 1$. The argument is proved.

Example 1.17 *Prove that the equality $\left(\dfrac{a}{(a,c)}, \dfrac{b}{(b,a)}, \dfrac{c}{(c,b)} \right) = 1$ holds.*

Proof From

$$\frac{a}{(a,c)} \,\bigg|\, \frac{a}{(a,b,c)}, \quad \frac{b}{(a,b)} \,\bigg|\, \frac{b}{(a,b,c)}, \quad \frac{c}{(b,c)} \,\bigg|\, \frac{c}{(a,b,c)},$$

we get

$$\left(\frac{a}{(a,c)}, \frac{b}{(b,a)}, \frac{c}{(c,b)} \right) \,\bigg|\, \left(\frac{a}{(a,b,c)}, \frac{b}{(a,b,c)}, \frac{c}{(a,b,c)} \right) = 1.$$

Therefore,

$$\left(\frac{a}{(a,c)}, \frac{b}{(b,a)}, \frac{c}{(c,b)} \right) = 1.$$

The proof is complete.

Definition 1.18 Let a_1, a_2 be two nonzero integers. If $a_1|l, a_2|l$, then l is said to be a common multiple of a_1 and a_2. More generally, let $a_1, \cdots, a_k$ be k nonzero integers. If $a_1|l, \cdots, a_k|l$, then l is said to be a common multiple of $a_1, \cdots, a_k$. $\square$

Definition 1.19 Let a_1, a_2 be two nonzero integers and l be a positive common multiple of a_1 and a_2. If, for any common multiple l' of a_1 and a_2, $l|l'$ holds, then l is called the least common multiple of a_1 and a_2, and is denoted by $[a_1, a_2]$. More generally, if l is a positive common multiple of $a_1, \cdots, a_k$ and for any common multiple l' of $a_1, \cdots, a_k$, $l|l'$ holds, then l is called the least common multiple of $a_1, \cdots, a_k$ and is denoted by $[a_1, \cdots, a_k]$. $\square$

We have the following results concerning the greatest common divisors and the least common multiples:

Theorem 1.20

1. *If $a_2|a_1$, then $[a_1, a_2] = |a_1|$; if $a_j|a_1$, $2 \leq j \leq k$, then $[a_1, a_2, \cdots, a_k] = |a_1|$.*

2. *For any $d|a_1$, $[a_1, a_2] = [a_1, a_2, d]$.*

3. *If $m > 0$, then $[ma_1, ma_2, \cdots, ma_k] = m[a_1, a_2, \cdots, a_k]$.*

4. *$(a_1, a_2, a_3, \cdots, a_k) = ((a_1, a_2), a_3, \cdots, a_k)$.*

5. *$(a_1, \cdots, a_{k+r}) = ((a_1, \cdots, a_k), (a_{k+1}, \cdots, a_{k+r}))$.*

Proof Let

$$L = [ma_1, ma_2, \cdots, ma_k], \quad L' = [a_1, a_2, \cdots, a_k].$$

From $ma_j|L, 1 \leq j \leq k$, we get $a_j|L/m, 1 \leq j \leq k$. Therefore, we have $L' \leq L/m$ by the definition of least common multiples. From $a_j|L'$, $1 \leq j \leq k$, we get $ma_j|mL', 1 \leq j \leq k$, and again we have $mL' \geq L$ by the definition of the least common multiple. These imply part 3 of the theorem.

Let

$$d = (a_1, a_2, a_3, \cdots, a_k), d' = ((a_1, a_2), a_3, \cdots, a_k).$$

We will prove $d = d'$. Since $d|a_j (1 \leq j \leq k)$, then $d|(a_1, a_2)$, $d|a_j$ $(3 \leq j \leq k)$; therefore, $d|d'$. Conversely, the facts $d'|(a_1, a_2)$,

$d'|a_j$ $(3 \leq j \leq k)$ imply $d|a_j$ $(1 \leq j \leq k)$, and hence, $d'|d$. This proves part 4 of the theorem.

Parts 1, 2, and 5 of the theorem are quite simple, so we leave the detailed proof to the reader.

We have further results concerning the greatest common divisors and the least common multiples.

Theorem 1.21 *If* $(m, a) = 1$, *then* $(m, ab) = (m, b)$.

Proof If $m = 0$, $a = \pm 1$, and so the result is obvious. If $m \neq 0$,

$$(m, b) = (m, b(m, a)) = (m, (mb, ab)) = (m, mb, ab) = (m, ab).$$

The theorem is proved.

Theorem 1.22 *Let* $(m, a) = 1$. *If* $m|ab$, *then* $m|b$.

Proof We have $|m| = (m, ab) = (m, b)$ from Theorem 1.21; therefore, $m|b$.

Example 1.23 *Let* $a > 2$ *be an odd number. Prove that*

1. *There must be a positive integer* $d \leq a - 1$, *such that* $a|2^d - 1$.

2. *If* d_0 *is the smallest number that satisfies* $a|2^{d_0} - 1$, *then* $a|2^h - 1$ *(* $h \in \mathbb{N}$ *) if and only if* $d_0|h$.

3. *There must be a positive integer* d *such that* $(2^d - 3, a) = 1$.

Proof
1. Consider a integers

$$2^0, \ 2^1, \ 2^2, \cdots, \ 2^{a-1}.$$

From the fact that $a \nmid 2^j$ $(0 \leq j < a)$ and the division algorithm, for every j, $0 \leq j < a$, there are q_j, r_j such that

$$2^j = q_j a + r_j, 0 < r_j < a.$$

There are at most $a - 1$ distinct values among the a remainders $r_0, r_1, \cdots, r_{a-1}$. By the pigeonhole principle, two of the remainders must be the same. We may assume $r_i = r_k$ for $0 \leq i < k < a$. Therefore,

$$a(q_k - q_i) = 2^k - 2^i = 2^i(2^{k-i} - 1).$$

Since $(a, 2) = 1$, we have $a|2^{k-i} - 1$. The argument is proved if we take $d = k - i \leq a - 1$.

2. As the sufficiency is obvious, we only need to show the necessity. By the division algorithm, we have

$$h = qd_0 + r, \ 0 \leq r < d_0.$$

Therefore,

$$2^h - 1 = 2^{qd_0+r} - 2^r + 2^r - 1 = 2^r(2^{qd_0} - 1) + (2^r - 1).$$

From $a|2^h - 1$ and $a|2^{qd_0} - 1$, it is easy to see that $a|2^r - 1$. This, together with the minimality of d_0, implies that $r = 0$, or $d_0|h$.

3. From 1, there exists d such that $a|2^d - 1$. This implies that

$$(2^d - 3, a) = (2^d - 1 - 2, a) = (-2, a) = 1.$$

Theorem 1.24 $[a_1, a_2](a_1, a_2) = |a_1 a_2|$.

Proof We first prove that when $(a_1, a_2) = 1$, the result is true. Let $l = [a_1, a_2]$; then $l|a_1 a_2$. On the other hand, $a_1|l$ implies that $l = a_1 l'$. From $a_2|l = a_1 l'$, $(a_2, a_1) = 1$ and Theorem 1.22, we get $a_2|l'$ and hence, $a_1 a_2|l$. The argument is proved.

Suppose that $(a_1, a_2) \neq 1$. From Theorem 1.13, we have

$$((a_1/(a_1, a_2), a_2/(a_1, a_2)) = 1;$$

hence,

$$\left[\frac{a_1}{(a_1, a_2)}, \frac{a_2}{(a_1, a_2)} \right] = \frac{|a_1 a_2|}{(a_1, a_2)^2}.$$

The proof is complete.

Example 1.25 *Let k be a positive integer. Prove that*

1. $(a^k, b^k) = (a, b)^k$.

2. *If* $(a, b) = 1$, $ab = c^k$, *then* $a = (a, c)^k$, $b = (b, c)^k$.

Proof From Theorem 1.13, we get

$$(a^k, b^k) = (a, b)^k \left(\frac{a^k}{(a,b)^k}, \frac{b^k}{(a,b)^k} \right) = (a, b)^k \left(\left(\frac{a}{(a,b)} \right)^k, \left(\frac{b}{(a,b)} \right)^k \right).$$

Since

$$\left(\left(\frac{a}{(a,b)} \right), \left(\frac{b}{(a,b)} \right) \right) = 1,$$

by Theorem 1.21, we have

$$\left(\left(\frac{a}{(a,b)} \right)^k, \left(\frac{b}{(a,b)} \right)^k \right) = 1.$$

This proves part 1.

Then $(a, b) = 1$ implies that $(a^{k-1}, b) = 1$; hence,

$$a = a(a^{k-1}, b) = (a^k, ab) = (a^k, c^k) = (a, c)^k.$$

Similarly, we have $b = (b, c)^k$.

Example 1.26 *Let p be a prime number. Prove that $\sqrt{p}$ is not a rational number.*

Proof Assume that there are positive integers a, b with $(a, b) = 1$ such that $\frac{a}{b} = \sqrt{p}$. This means that $\frac{a^2}{b^2} = p$, or $pb^2 = a^2$. Thus, $p|a$, $p^2|a^2$. This, in turn, forces that $p|b$. We get a contradiction to the fact that $(a, b) = 1$.

1.3   THE EUCLIDEAN ALGORITHM

The Euclidean algorithm is a fundamental algorithm in the field of mathematics. Important applications of this idea and technique have been found in many branches of mathematics. Using the Euclidean algorithm, one can find the greatest common divisor of finitely many integers. One can solve a linear Diophantine equation directly by applying this algorithm. The Euclidean algorithm is also useful in cryptography in several ways, for example, in breaking some encryption schemes or in analyzing the security of some encryption algorithms.

Theorem 1.27 (*The Euclidean algorithm*) *Let* a, b *be two integers with* $b \neq 0$ *and* $b \nmid a$. *Using the division algorithm repeatedly, we get the following* $k + 2$ *identities:*

$$
\begin{aligned}
a &= q_0 b + r_0, & 0 &< r_0 < |b|, \\
b &= q_1 r_0 + r_1, & 0 &< r_1 < r_0, \\
r_0 &= q_2 r_1 + r_2, & 0 &< r_2 < r_1, \\
&\cdots & &\cdots \\
r_{k-3} &= q_{k-1} r_{k-2} + r_{k-1}, & 0 &< r_{k-1} < r_{k-2}, \\
r_{k-2} &= q_k r_{k-1} + r_k, & 0 &< r_k < r_{k-1}, \\
r_{k-1} &= q_{k+1} r_k.
\end{aligned}
$$

Proof Applying the division algorithm to a, b, we get the first relation, as a is not divisible by b. Similarly, if b is not divisible by r_0, then we get the second relation. Repeating this procedure, we have

$$
|b| > r_0 > r_1 > \cdots > r_{j-1} > 0,
$$

and the first $j - 2$ relations hold. If $r_{j-1} | r_{j-2}$, then the theorem is proved by taking $k = j - 2$; if $r_{j-1} \nmid r_{j-2}$, then we apply the division algorithm to r_{j-2}, r_{j-1}. Since there are only finitely many positive integers that are less than $|b|$ and any integer is divisible by 1, there must exist some k such that either $1 < r_k | r_{k-1}$ or $1 = r_k | r_{k-1}$. The theorem is finally proved.

The Euclidean algorithm described in Theorem 1.27 is about the smallest positive remainders. In Section 10.1 of this book, we will present the extended Euclidean algorithm and an estimation of its time complexity.

Theorem 1.28 *Under the same conditions and with the same notations as in Theorem 1.27, we have*

1. $r_k = (a, b)$.

2. *There exist* x_0, x_1 *such that* $(a, b) = a x_0 + b x_1$.

Proof

1. Starting from the last relation of Theorem 1.27, and tracing back step by step, we get

$$
r_k = (r_k, r_{k-1}) = \cdots = (r_1, r_0) = (r_0, b) = (a, b).
$$

   The argument follows.

2. From the $(k+1)$st relation of the Euclidean algorithm, (a,b) can be represented as an integer linear combination of r_{k-1} and r_{k-2}, the term r_{k-1} can be replaced by using the kth relation, and (a,b) becomes an integer linear combination of r_{k-2} and r_{k-3}. Repeating the procedure using the ith relation for $i = k-1, \cdots,$ $2, 1$, we see that (a,b) is an integer linear combination of a and b.

Theorem 1.28 gives a very convenient concrete algorithm for computing the greatest common divisors; it also contains a procedure for finding x_1, x_0. The following is a corollary of Theorem 1.28.

Corollary 1.29 *Let $a_1, \cdots, a_k$ be integers that are not all zero. Then there must be integers $x_{1,0}, \cdots, x_{k,0}$ such that*

$$(a_1, \cdots, a_k) = a_1 x_{1,0} + \cdots + a_k x_{k,0}.$$

Example 1.30 *Find the greatest common divisor of 42823 and 6409, and represent it as an integer linear combination of 42823 and 6409.*

Solution: From the Euclidean algorithm,

$$42823 = 6 \cdot 6409 + 4369$$
$$6409 = 1 \cdot 4369 + 2040$$
$$4369 = 2 \cdot 2040 + 289$$
$$2040 = 7 \cdot 289 + 17$$
$$289 = 17 \cdot 17$$

we see that $(42823, 6049) = 17$.

Reversing the above process, we see that

$$17 = 2040 - 7 \cdot 289$$
$$17 = -7 \cdot 4369 + 15 \cdot 2040$$
$$17 = 15 \cdot 6409 - 22 \cdot 4369$$
$$17 = -22 \cdot 42823 + 147 \cdot 6409.$$

This gives the linear combination

$$(42823, 6409) = -22 \cdot 42823 + 147 \cdot 6409.$$

Example 1.31 *If $(a,b) = 1$, any integer n is an integer linear combination of a, b.*

Proof Since $(a, b) = 1$, there exist x_0, y_0 such that $ax_0 + by_0 = 1$. So we can take $x = nx_0, y = ny_0$ and get

$$n = xa + yb.$$

Example 1.32 *Assume that a, m, n are positive integers and $a \geq 2$. Prove that*

$$(a^m - 1, a^n - 1) = a^{(m,n)} - 1.$$

Proof We may assume that $m \geq n$. By the division algorithm,

$$m = q_1 n + r_1, \qquad 0 \leq r_1 < n.$$

So we have

$$a^m - 1 = a^{q_1 n + r_1} - a^{r_1} + a^{r_1} - 1 = a^{r_1}(a^{q_1 n} - 1) + a^{r_1} - 1.$$

This, together with $a^n - 1 | a^{q_1 n} - 1$, implies that

$$(a^m - 1, a^n - 1) = (a^{r_1} - 1, a^n - 1).$$

Notice that $(m, n) = (n, r_1)$, we have $(m, n) = n$ when $r_1 = 0$, and the result holds. If $r_1 > 0$, then we can make the same argument to $(a^{r_1} - 1, a^n - 1)$ and get the result by using the Euclidean algorithm. The proof is complete.

Example 1.33 *The Fibonacci sequence is defined by*

$$f_1 = 1, f_2 = 1, f_n = f_{n-1} + f_{n-2}, n \geq 3.$$

Prove that any two consecutive Fibonacci numbers are coprime.

Proof Obviously, we have $(f_1, f_2) = 1, (f_2, f_3) = 1$. Now let us prove that $(f_k, f_{k+1}) = 1, k \geq 3$. By the definition of the Fibonacci sequence, we can perform the Euclidean algorithm as follows:

$$f_{k+1} = f_k \cdot 1 + f_{k-1},$$
$$f_k = f_{k-1} \cdot 1 + f_{k-2},$$
$$\cdots$$
$$f_5 = f_4 \cdot 1 + f_3,$$
$$f_4 = f_3 \cdot 1 + f_2,$$
$$f_3 = f_2 \cdot 2.$$

Therefore, $(f_k, f_{k+1}) = f_2 = 1$. The proof is complete.

1.4   SOLVING LINEAR DIOPHANTINE EQUATIONS

The Euclidean algorithm has a wider range of applications. Here we introduce one of them—solving the linear Diophantine equations. The general form of a linear Diophantine equation is

$$a_1 x_1 + \cdots + a_k x_k = c, \tag{1.1}$$

where the integer $k \geq 2$; $c, a_1, \cdots, a_k$ are also integers and $a_1, \cdots, a_k$ are not all zero; and $x_1, \cdots, x_k$ are integer variables.

We first present a sufficient and necessary condition under which Equation 1.1 has a solution.

Theorem 1.34 *The Diophantine Equation 1.1 has a solution if and only if $(a_1, \cdots, a_k)|c$. If Equation 1.1 has solutions, then the set of its solutions is the same as that of the following Diophantine equation:*

$$\frac{a_1}{d} x_1 + \cdots + \frac{a_k}{d} x_k = \frac{c}{d}, \tag{1.2}$$

where $d = (a_1, \cdots, a_k)$.

Proof   As the necessity is obvious, we only need to show the sufficiency. If $d|c$, say $c = dc_1$, then there must be integers $y_{1,0}, \cdots, y_{k,0}$ such that

$$a_1 y_{1,0} + \cdots + a_k y_{k,0} = d.$$

Therefore, $x_1 = c_1 y_{1,0}, \cdots, x_k = c_1 y_{k,0}$ is a solution to Equation 1.1 and the sufficiency is proved.

If Equation 1.1 has a solution, then $d|c$ must hold. In this case, Equations 1.1 and 1.2 represent the same equation, and thus we have proved the last argument.

Theorem 1.35 *Assume that the linear Diophantine equation in two variables*

$$a_1 x_1 + a_2 x_2 = c \tag{1.3}$$

has a solution $x_{1,0}, x_{2,0}$. Then all of its solutions are

$$\begin{cases} x_1 = x_{1,0} + \dfrac{a_2}{(a_1, a_2)} t, \\[2mm] x_2 = x_{2,0} - \dfrac{a_1}{(a_1, a_2)} t, \end{cases}$$

where $t = 0, \pm 1, \pm 2, \cdots$.

Proof It is easy to verify that every pair of x_1, x_2 given in the theorem is a solution of Equation 1.3. Conversely, let x_1, x_2 be a solution of Equation 1.3; then we have

$$a_1 x_1 + a_2 x_2 = a_1 x_{1,0} + a_2 x_{2,0}.$$

This yields

$$a_1 (x_1 - x_{1,0}) = -a_2 (x_2 - x_{2,0}),$$

$$\frac{a_1}{(a_1, a_2)} (x_1 - x_{1,0}) = -\frac{a_2}{(a_1, a_2)} (x_2 - x_{2,0}).$$

Since $\left(\dfrac{a_1}{(a_1, a_2)}, \dfrac{a_2}{(a_1, a_2)} \right) = 1$, we have

$$x_1 - x_{1,0} = \frac{a_2}{(a_1, a_2)} t, \quad x_2 - x_{2,0} = -\frac{a_1}{(a_1, a_2)} t.$$

The proof is complete.

From this theorem, we can summarize the steps of solving the linear Diophantine equation in two variables:

1. Determine whether $(a_1, a_2) | c$ holds.

2. If $(a_1, a_2) | c$ holds, then the equation has a solution. Try to find a specific solution $x_{1,0}, x_{2,0}$, and get all solutions by using the above formula.

Let us use an example to explain the steps.

Example 1.36 *Solve the linear Diophantine equation in two variables* $3x_1 + 5x_2 = 11$.

Solution: We can easily get a specific solution $x_1 = 2$, $x_2 = 1$ of this equation. Since $(3, 5) = 1$, the solutions of the equation are

$$\begin{cases} x_1 = 2 + 5t, \\ x_2 = 1 - 3t, \end{cases}$$

where $t = 0, \pm 1, \pm 2, \cdots$.

1.5 PRIME FACTORIZATION OF INTEGERS

In theory, any integer can be written as a product of primes. However, prime factorization of a big integer is a difficult problem. So far, a deterministic polynomial time algorithm for factoring a big integer has not yet been found. However, the hardness of integer factorization sets up a theoretical foundation of security for some public key systems. Now let us introduce some theorems that relate to the prime factorization of integers.

Lemma 1.37 *If p is a prime and $p|a_1a_2$, then at least one of $p|a_1$ and $p|a_2$ holds. In general, if $p|a_1\cdots a_k$, then at least one of $p|a_1,\cdots,p|a_k$ holds.*

Theorem 1.38 (*The Fundamental Theorem of Arithmetic*) *Let $a > 1$. Then we have*

$$a = p_1 p_2 \cdots p_s, \tag{1.4}$$

where $p_j(1 \leq j \leq s)$ are primes and, up to ordering of the factors, Equation 1.4 is unique.

Proof Existence: We prove this by mathematical induction. If $a = 2$, the result is true as 2 is a prime. Assume that the result is true for $2 \leq a < n$. Consider the case that $a = n$. The result is obviously true if n is prime. If n is composite, then $n = n_1 n_2$ for $2 \leq n_1$, $n_2 < n$. By our induction hypothesis, n_1, n_2 are both products of prime numbers:

$$n_1 = p_{11} \cdots p_{1s}, \quad n_2 = p_{21} \cdots p_{2r}.$$

Therefore, a is a product of primes

$$a = n = n_1 n_2 = p_{11} \cdots p_{1s} p_{21} \cdots p_{2r}.$$

That proves the existence of prime factorization of integers.

Uniqueness: If there are two forms of prime factorization

$$a = p_1 p_2 \cdots p_s, \quad p_1 \leq p_2 \leq \cdots \leq p_s,$$

$$a = q_1 q_2 \cdots q_r, \quad q_1 \leq q_2 \leq \cdots \leq q_r,$$

where $p_i(1 \leq i \leq s), q_i(1 \leq i \leq r)$ are primes. We will prove that $r = s$ and $p_i = q_i, (1 \leq i \leq s)$. Assume that $r \geq s$. Since $q_1|a = p_1 p_2 \cdots p_s$, there must be a p_j such that $q_1|p_j$. Notice that q_1 and p_j are both primes, so $q_1 = p_j$. Similarly, since $p_1|a = q_1q_2\cdots q_r$, there must be a q_i such that $p_1|q_i$; hence, $p_1 = q_i$. But $q_1 \leq q_i = p_1 \leq p_j$, so we have $p_1 = q_1$. This also implies

$$q_2q_3\cdots q_r = p_2p_3\cdots p_s.$$

Repeating the same process, we have

$$q_2 = p_2, \cdots, q_s = p_s, q_{s+1} = 1, \cdots, q_r = 1.$$

There will be no terms $q_{s+1}, \cdots q_r$, namely, $r = s$. The theorem is proved.

Corollary 1.39 *If $a > 1$, then we have*

$$a = p_1^{\alpha_1} \cdots p_s^{\alpha_s}, \quad p_1 < p_2 < \cdots < p_s. \tag{1.5}$$

Equation 1.5 is called the canonical prime factorization of a.

The proof is trivial. We only need to make some rearrangements of the primes in Equation 1.4.

Corollary 1.40 *If $a = p_1^{\alpha_1} \cdots p_s^{\alpha_s}$, and $p_i(1 \leq i \leq s)$ are distinct primes, then d is a positive factor of a if and only if*

$$d = p_1^{e_1} \cdots p_s^{e_s}, \quad 0 \leq e_j \leq \alpha_j, \quad 1 \leq j \leq s.$$

Corollary 1.41 *Let $p_1 < p_2 < \cdots < p_s$ be primes and*

$$a = p_1^{\alpha_1} \cdots p_s^{\alpha_s},$$

$$b = p_1^{\beta_1} \cdots p_s^{\beta_s}.$$

If some α_j or β_j are allowed to be zero, then

$$(a, b) = p_1^{\delta_1} \cdots p_s^{\delta_s}, \quad \delta_j = \min(\alpha_j, \beta_j), \quad 1 \leq j \leq s,$$

$$[a, b] = p_1^{\gamma_1} \cdots p_s^{\gamma_s}, \quad \gamma_j = \max(\alpha_j, \beta_j), \quad 1 \leq j \leq s,$$

and

$$(a, b)[a, b] = ab.$$

Example 1.42 *Prove that* $(a, [b, c]) = [(a, b), (a, c)]$.

Proof If one of a, b, c is 0, the equality holds. So we assume that a, b, c are positive integers. Let $p_1 < p_2 < \cdots < p_s$ be primes such that

$$a = p_1^{\alpha_1} \cdots p_s^{\alpha_s}, \quad b = p_1^{\beta_1} \cdots p_s^{\beta_s}, \quad c = p_1^{\gamma_1} \cdots p_s^{\gamma_s}.$$

From Corollary 1.41, we have

$$(a, [b, c]) = p_1^{\eta_1} \cdots p_s^{\eta_s},$$

$$\eta_j = \min(\alpha_j, \max(\beta_j, \gamma_j)), \quad 1 \leq j \leq s;$$

$$[(a, b), (a, c)] = p_1^{\tau_1} \cdots p_s^{\tau_s},$$

$$\tau_j = \max(\min(\alpha_j, \beta_j), \min(\alpha_j, \gamma_j)), \quad 1 \leq j \leq s.$$

It is easy to verify that $\tau_j = \eta_j (1 \leq j \leq s)$ always holds regardless of the magnitude of α_j, β_j, and γ_j. The result is proved.

Corollary 1.43 *Let a be a positive integer and $\tau(a)$ denote the number of positive factors of a (usually called the divisor function). If a has the canonical prime factorization (Equation 1.5), then*

$$\tau(a) = (\alpha_1 + 1) \cdots (\alpha_s + 1) = \tau(p_1^{\alpha_1}) \cdots \tau(p_s^{\alpha_s}).$$

Corollary 1.44 *Let integer $a \geq 2$.*

1. *If a is a composite, then there is a prime $p|a$ and $p \leq a^{\frac{1}{2}}$.*

2. *If a is represented as Equation 1.4, then there is a prime $p|a$ and $p \leq a^{\frac{1}{s}}$.*

Remark Corollary 1.44 gives a method for finding prime—the Sieve of Eratosthenes. The Sieve of Eratosthenes is used to list all primes that are less than or equal to a positive integer n. This method can be described as follows:

First, starting from 2, list all positive integers that are less than or equal to n in ascending order:

$$2, 3, 4, 5, 6, 7, 8, 9, 10, 11, 12, 13, 14, 15, 16, 17, 18, 19, 20, 21, \cdots, n.$$

Starting from the first prime number 2, delete all multiples $2m$ of 2 from the list, where the positive integer m satisfies $2 < 2m \leq n$. The first sequence is created:

$$2, 3, 5, 7, 9, 11, 13, 15, 17, 19, 21, \cdots, n.$$

Second, starting from the second prime number 3, delete all multiples $3m$ of 3 from the first sequence, where the positive integer m satisfies $3 < 3m \le n$. We get the second sequence:

$$2, 3, 5, 7, 11, 13, 17, 19, \cdots, n.$$

Performing a similar procedure repeatedly, we can get all primes that are less than or equal to n in finite steps. In fact, when step i is reached, if the ith prime p is larger than $\sqrt{n}$, then the sequence produced in step $i-1$ is already the sequence of primes that are not larger than n.

Finally, we will give a computation formula for the prime factorization of $n!$. Before discussing the prime factorization of $n!$, however, let us take a look at a related arithmetic function $[x]$.

Definition 1.45 Let x be a real number and $[x]$ denote the largest integer that is less than or equal to x which is called the integer part of x. Namely, $[x]$ is a positive integer that satisfies $[x] \le x < [x] + 1$. The number $\{x\} = x - [x]$ is called the decimal part of x. It is obvious that $0 \le \{x\} < 1$, and x is an integer if and only if $\{x\} = 0$. □

The arithmetic function $[x]$ has the following properties.

Theorem 1.46 *Let x, y be real numbers. We have*

1. *If $x \le y$, then $[x] \le [y]$.*

2. *If $x = m + v$, where m is an integer and $0 \le v < 1$, then $m = [x]$, $v = \{x\}$. In particular, if $0 \le x < 1$, then $[x] = 0, \{x\} = x$.*

3. *For any integer m, we have $[x + m] = [x] + m, \{x + m\} = \{x\}$. $\{x\}$ is a periodic function with period 1.*

4. *$[x] + [y] \le [x + y] \le [x] + [y] + 1$ and exactly one equality holds.*

5. $[-x] = \begin{cases} -[x], & x \in \mathbb{Z}, \\ -[x] - 1, & x \notin \mathbb{Z}, \end{cases}$ $\{-x\} = \begin{cases} -\{x\} = 0, & x \in \mathbb{Z}, \\ 1 - \{x\}, & x \notin \mathbb{Z}. \end{cases}$

6. *For positive integer m, we have $\left[\dfrac{[x]}{m}\right] = \left[\dfrac{x}{m}\right]$.*

7. *The least integer that is not smaller than x is $-[-x]$.*

8. *Let a and N be positive integers. Then the number of positive integers in $1, 2, \cdots, N$ that are divisible by a is $[N/a]$.*

Proof

1. The result follows from the fact that $[x] \leq x \leq y < [y] + 1$.

2. The result follows from the fact that $m \leq x < m + 1$.

3. The result follows from the fact that $[x] + m \leq x + m < ([x] + m) + 1$.

4. Notice that $x + y = [x] + [y] + \{x\} + \{y\}$ and $0 \leq \{x\} + \{y\} < 2$. If $0 \leq \{x\} + \{y\} < 1$, then from statement 2, $[x+y] = [x] + [y]$. If $1 \leq \{x\} + \{y\} < 2$, then $x + y = ([x] + [y] + 1) + (\{x\} + \{y\} - 1)$. From statement 2, we get $[x+y] = [x] + [y] + 1$.

5. The result is obvious if x is an integer. If x is not an integer, $-x = -[x] - \{x\} = -[x] - 1 + 1 - \{x\}, 0 \leq -\{x\} + 1 < 1$, and the result follows by using statement 2.

6. From the division algorithm, there are q, r such that

$$[x] = qm + r, 0 \leq r < m,$$

   that is,

$$[x]/m = q + r/m, 0 \leq r/m < 1.$$

   Together with statement 2, this implies $[[x]/m] = q$. On the other hand,

$$x/m = [x]/m + \{x\}/m = q + (\{x\} + r)/m.$$

   Notice that $0 \leq (\{x\} + r)/m < 1$, so we have $[x/m] = q$ by using statement 2 and the result holds.

7. Let a be the least integer that is not smaller than x, that is, $a - 1 < x \leq a$, since $-a \leq -x < -a + 1$, so $-a = [-x]$. That is, $a = -[-x]$.

8. The numbers that are divisible by a are $a, 2a, 3a, \cdots$. Suppose that there are k integers in $1, 2, \cdots, N$ that are divisible by a; then we must have $ka \leq N < (k+1)a$, that is, $k \leq N/a < (k+1)$, so the result follows.

   We will need an additional symbol.

Definition 1.47 Let k be a nonnegative integer, and the symbol $a^k\|b$ denotes that b is exactly divisible by the kth power of a, that is, $a^k|b$, $a^{k+1} \nmid b$. □

Theorem 1.48 *Let n be a positive integer, and p a prime. Let $\alpha = \alpha(p,n)$ satisfy $p^\alpha\|n!$. Then*

$$\alpha = \alpha(p,n) = \sum_{j=1}^{\infty} \left[\frac{n}{p^j}\right].$$

Proof The equality in the theorem is in fact a finite sum because there is a k such that

$$p^k \le n < p^{k+1}.$$

So,

$$\alpha = \sum_{j=1}^{k} \left[\frac{n}{p^j}\right].$$

Let j be a given positive integer; c_j denotes the number of integers in $1, 2, \cdots, n$ that are divisible by p^j, and d_j denotes the number of integers in $1, 2, \cdots, n$ that are exactly divisible by p^j. It is easy to see that

$$d_j = c_j - c_{j+1}.$$

By statement 8 of Theorem 1.46,

$$d_j = [n/p^j] - [n/p^{j+1}].$$

It is easy to check that when $j > k$, $d_j = 0$.

Next, we divide $1, 2, \cdots, n$ into k disjoint sets with the jth set consisting of all integers in $1, 2, \cdots, n$ that are exactly divisible by p^j. Thus, the product of all numbers in the jth set can be exactly divisible by the $j \cdot d_j$ power of p, so we have

$$\alpha = 1 \cdot d_1 + 2 \cdot d_2 + \cdots + k \cdot d_k.$$

The theorem is proved.

Corollary 1.49 *Let n be a positive integer. We have*

$$n! = \prod_{p \leq n} p^{\alpha(p,n)},$$

where the product is taken over all primes less than or equal to n.

In addition if $p_2 < p_1$, then obviously we have $\alpha(p_1, n) \leq \alpha(p_2, n)$.

Example 1.50 *Find the number of trailing zeros in the decimal form of $80!$.*

Solution: This is equivalent to finding k such that $10^k \| 80!$. In the prime factorization of $80!$, the smaller the prime p, the larger is the number $\alpha = \alpha(p, n)$. So we only need to find the power of 5 in $80!$:

$$\alpha = \alpha(5, 80) = \sum_{j=1}^{\infty} \left[\frac{80}{5^j}\right] = \left[\frac{80}{5}\right] + \left[\frac{80}{25}\right] = 19.$$

This shows that the number of trailing zeros in $80!$ is 19.

Example 1.51 *Let $a_j > 0$, $1 \leq j \leq s$ be integers and $n = a_1 + a_2 + \cdots + a_s$. Prove that $\dfrac{n!}{a_1! a_2! \cdots a_s!}$ is an integer.*

Proof By Corollary 1.49, it suffices to show for any prime p,

$$\alpha(p, n) \geq \alpha(p, a_1) + \alpha(p, a_2) + \cdots + \alpha(p, a_s)$$

holds. This means that we need to show for any $j \geq 1$,

$$\left[\frac{n}{p^j}\right] \geq \left[\frac{a_1}{p^j}\right] + \left[\frac{a_2}{p^j}\right] + \cdots + \left[\frac{a_s}{p^j}\right].$$

By the fact that $n = a_1 + a_2 + \cdots + a_s$ and item 4 of Theorem 1.46, we know that the previous inequality holds. Hence, $\dfrac{n!}{a_1! a_2! \cdots a_s!}$ is an integer.

EXERCISES

1.1 Use the Sieve of Eratosthenes to find all primes within 200.

1.2 If a is an odd number, prove that there must be a positive integer d such that $(2^d - 5, a) = 1$.

1.3 Let a be a positive integer and $\sigma(a)$ denote the sum of all positive divisors of a. Prove that $\sigma(1) = 1$. If a has canonical prime factorization $a = p_1^{\alpha_1} p_2^{\alpha_2} \cdots p_s^{\alpha_s}$, $p_1 < p_2 < \cdots < p_s$, prove that

$$\sigma(a) = \frac{p_1^{\alpha_1+1} - 1}{p_1 - 1} \cdots \frac{p_s^{\alpha_s+1} - 1}{p_s - 1} = \prod_{j=1}^{s} \frac{p_j^{\alpha_j+1} - 1}{p_j - 1}$$

$$= \sigma(p_1^{\alpha_1}) \cdots \sigma(p_s^{\alpha_s}).$$

1.4 Find the greatest common divisor of 198 and 252, and represent it as an integer linear combination of 198 and 252.

1.5 Find least common multiples:

(1) $[220, 284]$.

(2) $[10773, 23446]$.

1.6 Find the solutions of $117x_1 + 21x_2 = 38$.

1.7 Prove that if $n > 1$, then $1 + 1/2 + \cdots + 1/n$ is not an integer.

1.8 Prove that

(1) The largest integer that is less than x is $-[-x] - 1$.

(2) The least integer that is greater than x is $[x] + 1$.

(3) The integers that are closest to x are $[x + 1/2]$ and $-[-x + 1/2]$. These two integers have the same distances to x if $x + 1/2$ is an integer; otherwise, these two numbers are the same.

1.9 Perform the following conversions:

(1) Write the decimal integer 21701 into radix-8 form.

(2) Write the decimal integer 65537 into radix-16 form.

1.10 Prove that for any positive integer a,

$$5 | a^5 - a.$$

1.11 From the division algorithm, for any given integers a, b with $a \neq 0$, there must be a pair of integers q and r such that

$$b = qa + r, \quad 0 \leq r < |a|.$$

Prove that $q = [\frac{b}{a}], r = b - a \cdot [\frac{b}{a}]$.

1.12 Prove that $13 | a^2 - 7b^2$ if and only if $13 | a$ and $13 | b$.

1.13 Find the number of trailing zeros in the decimal form of $123!$.

1.14 If p is a prime, the number $M_p = 2^p - 1$ is called a Mersenne number. Give a direct proof of the fact that all Mersenne numbers are pairwise coprime, by representing this family of integers in binary form and applying the Euclidean algorithm (with all numbers appeared in the algorithm in binary form).

1.15 For the polynomial with integer coefficients,

$$p(x) = a_n x^n + a_{n-1} x^{n-1} + \cdots + a_0, a_n \neq 0.$$

Prove that there are infinitely many integer values of x such that $p(x)$ is a composite.

CHAPTER 2

Congruences

THE MAIN subject of this chapter is the basic concepts and properties of the congruence theory. In addition to introducing related concepts such as congruences, congruence relations, congruence classes, complete system of residues, and reduced system of residues, we describe several methods of constructing a complete system of residues and a reduced system of residues. Applications of Euler's theorem in algorithms for cryptography are also discussed.

2.1 CONGRUENCES

Definition 2.1 For a given positive integer m, if $m|a - b$, namely, there is a integer k such that $a - b = km$, then a is said to be congruent to b modulo m, and b is a residue of a with respect to the modulus m, denoted by

$$a \equiv b \pmod{m}. \tag{2.1}$$

Otherwise, a is said to be not congruent to b modulo m, or b is not a residue of a with respect to the modulus m, denoted by

$$a \not\equiv b \pmod{m}.$$

Relation 2.1 is called a congruence relation with respect to the modulus m, or simply a congruence relation. If $0 \leq b < m$, then b is called the smallest nonnegative residue of a with respect to the modulus m; if $1 \leq b \leq m$, then b is called the smallest positive residue of a with respect to the modulus m; if $-m/2 < b \leq m/2$ (or $-m/2 \leq b < m/2$), then b is called the smallest absolute residue of a with respect to the modulus m. □

Theorem 2.2 *a is congruent to b modulo m if and only if the smallest nonnegative remainders of a and b divided by m are the same; that is, if*

$$a = q_1 m + r_1, \qquad 0 \le r_1 < m;$$
$$b = q_2 m + r_2, \qquad 0 \le r_2 < m,$$

then $r_1 = r_2$.

Proof From $a - b = (q_1 - q_2)m + (r_1 - r_2)$, we see that $m | a - b$ if and only if $m | r_1 - r_2$. This, together with $0 \le |r_1 - r_2| < m$, yields $r_1 = r_2$.

By Theorem 2.2, checking whether the smallest nonnegative remainders are the same or not can be used to define congruence. For a fixed modulus m, the following simple properties are immediate from the definition of congruence relations.

Property 1 Congruence is an equivalence relation. It satisfies

1. Reflexivity: $a \equiv a \pmod{m}$.

2. Symmetry: $a \equiv b \pmod{m} \Leftrightarrow b \equiv a \pmod{m}$.

3. Transitivity: $a \equiv b \pmod{m}$, $b \equiv c \pmod{m} \Rightarrow a \equiv c \pmod{m}$.

Property 2 Addition and subtraction can be applied to congruence relations. If

$$a \equiv b \pmod{m}, \quad c \equiv d \pmod{m},$$

then

$$a \pm c \equiv b \pm d \pmod{m}.$$

Property 3 Congruence relations can be multiplied. If

$$a \equiv b \pmod{m}, \quad c \equiv d \pmod{m},$$

then

$$ac \equiv bd \pmod{m}.$$

Based on these properties, we have the following definition.

Definition 2.3 Let $f(x) = a_n x^n + \cdots + a_0$, $g(x) = b_n x^n + \cdots + b_0$ be two polynomials of integer coefficients with

$$a_j \equiv b_j \pmod{m}, \text{ for } 0 \le j \le n.$$

Then the polynomial $f(x)$ is said to be congruent to the polynomial $g(x)$ modulo m, denoted by $f(x) \equiv g(x) \pmod{m}$. □

It is obvious that if $a \equiv b \pmod{m}$, then

$$f(a) \equiv g(b) \pmod{m}.$$

Property 4 If $d \geq 1$, $d|m$, and the congruence relation (2.1) holds, then

$$a \equiv b \pmod{d}.$$

Property 5 The congruence relation

$$ca \equiv cb \pmod{m}$$

is equivalent to

$$a \equiv b \pmod{m/(c,m)}.$$

In particular, if $(c,m) = 1$, then $a \equiv b \pmod{m}$.

Proof From the properties of congruence relations,

$$ca \equiv cb \pmod{m} \Leftrightarrow \frac{m}{(c,m)} \left| \frac{c}{(c,m)}(a-b). \right.$$

Since $(m/(c,m), c/(c,m)) = 1$, we have

$$ca \equiv cb \pmod{m} \Leftrightarrow \frac{m}{(c,m)} \left| a-b. \right.$$

The proof is complete.

From property 5, congruence relations satisfy the law of cancellation as long as $(c,m) = 1$.

Property 6 If $m \geq 1$, $(a,m) = 1$, then there is a c such that

$$ca \equiv 1 \pmod{m}.$$

We call c the inverse of a modulo m, and denote it by $a^{-1}(\bmod\, m)$.

Proof By the Euclidean algorithm, there exist x_0, y_0 such that $ax_0 + my_0 = 1$. The requirement is satisfied if we take $c = x_0$.

Property 6 provides an efficient method for finding $a^{-1} \pmod{m}$. This is yet another important application of the Euclidean algorithm.

Example 2.4 *Find inverses modulo $p = 11$ of all elements that are coprime to* 11.

Solution: A simple calculation produces the following table:

a	1	2	3	4	5	6	7	8	9	10
a^{-1}	1	6	4	3	9	2	8	7	5	10

Property 7 The congruence relations

$$a \equiv b \pmod{m_j}, \; j = 1, 2, \cdots, k,$$

hold simultaneously if and only if $a \equiv b \pmod{[m_1, \cdots, m_k]}$.

Proof Since a common multiple must be a multiple of the least common multiple, we see that

$$m_j | a - b, \; j = 1, \cdots, k \Leftrightarrow [m_1, \cdots, m_k] | a - b.$$

Hence, the result holds.

The next example provides a criterion of determining whether a number is divisible by 9 by using the congruence operation.

Example 2.5 *Let n be an integer. Find a sufficient and necessary condition of n being divisible by 9.*

Solution: Let $n = a_0 10^k + \cdots + a_{k-1} 10 + a_k$. Since $10^i \equiv 1 \pmod 9$, $1 \le i \le k$, by the properties 2 and 3, we get

$$n \equiv a_0 + a_1 + \cdots + a_k \pmod 9.$$

So

$$9 | a_0 + a_1 + \cdots + a_k \Leftrightarrow 9 | n.$$

Similar to the argument of Example 2.5, we can get n is divisible by 3 if and only if $3 | a_0 + a_1 + \cdots + a_k$. This has been a well-known criterion of determining whether a number is divisible by 3.

Example 2.6 *Find $6^{125} \pmod{41}$.*

Solution: Since

$$6^2 \equiv -5 \pmod{41}; \quad 6^4 \equiv 25 \pmod{41}; \quad 6^5 \equiv 27 \pmod{41};$$
$$6^{10} \equiv -9 \pmod{41}; \quad 6^{20} \equiv -1 \pmod{41}; \quad 6^{40} \equiv 1 \pmod{41}.$$

Therefore, $6^{125} = 6^{40 \times 3 + 5} \equiv 27 \pmod{41}$.

Remark Computing the power $a^x \pmod m$ modulo m is an important algorithm in cryptography. We will discuss it in more detail in Chapter 10.

2.2 RESIDUE CLASSES AND SYSTEMS OF RESIDUES

Congruence is an equivalence relation, so we can partition integers according to congruence relations. In this section, we will describe some concepts and special properties related to such equivalence classes.

Definition 2.7 Given a positive integer m, the set of all integers that are congruent modulo m is called a residue class (congruence class) modulo m. We use r (mod m) to denote the residue class modulo m where r belongs. If $(r, m) = 1$, the residue class r (mod m) is called a reduced (or coprime) residue class. The number of all reduced residue classes is denoted by $\varphi(m)$ and is usually called the Euler function. □

It is obvious that $\varphi(m)$ is the number of positive integers that are smaller than m and coprime to m. For residue classes, we have the following properties:

Property 8 For a given m, there are m and only m distinct residue classes modulo m, and the following are satisfied:

1. $\mathbb{Z} = \bigcup\limits_{r=0}^{m-1} r \text{ (mod } m).$

2. $i \text{ (mod } m) \bigcap j \text{ (mod } m) = \emptyset, \ 0 \le i, j < m, \ i \ne j.$

Property 8 is another description of partitioning all integers with respect to the least nonnegative remainders discussed in Section 1.1. Usually, 0 (mod m), $\cdots, (m-1)$ (mod m) are also denoted by $\overline{0}, \cdots, \overline{m-1}$.

Definition 2.8 For each residue class $\overline{i}$ modulo m, we take an arbitrary $a_j \in \overline{i}$, $0 \le i < m$ and call $a_0, a_1, \cdots, a_{m-1}$ a complete system of residues modulo m, denoted by $\mathbb{Z}_m$. Usually, we call $0, 1, 2, \cdots, m-1$ the least nonnegative (complete) system of residues modulo m; $1, 2, 3, \cdots, m-1, m$ the least positive system of residues modulo m; and $-[\frac{m}{2}], -\frac{m}{2} + 1, \cdots, 0, 1, \cdots, [\frac{m}{2} - 1]$ or $-[\frac{m}{2}] + 1, -[\frac{m}{2}] + 2, \cdots, 0, 1, \cdots, [\frac{m}{2}]$ the least absolute system of residues modulo m. □

Obviously, if $a_0, a_1, \cdots, a_{m-1}$ is a complete system of residues, then for any $a \in \mathbb{Z}$, there is one and only one $a_i, 0 \le i < m-1$ that is congruent to a modulo m.

Definition 2.9 For each reduced residue class $\overline{k}_j$ modulo m, we take an arbitrary $a_i \in \overline{k}_i$, $0 \le i < \varphi(m)$, $0 \le k_i < m$, $(k_i, m) = 1$, and call $a_0, a_1, \cdots, a_{\varphi(m)-1}$ a reduced system of residues modulo m, and denote this as $\mathbb{Z}_m^*$. □

Obviously, if $a_0, a_1, \cdots, a_{\varphi(m)-1}$ is a reduced system of residues, then for any $a \in \mathbb{Z}$, $(a, m) = 1$, there is one and only one a_i, $(0 \le i < \varphi(m) - 1)$ that is congruent to a modulo m.

The next theorem provides a common way of determining whether a set is a complete system of residues or a reduced system of residues modulo m.

Theorem 2.10

1. *m integers form a complete system of residues modulo m if and only if these m integers are mutually incongruent modulo m.*

2. *$\varphi(m)$ integers form a reduced system of residues modulo m if and only if these $\varphi(m)$ integers are mutually incongruent modulo m and each of them is coprime to m.*

Proof

1. Suppose that $y_0, \cdots, y_{m-1}$ are mutually incongruent modulo m, then, $y_0, \cdots, y_{m-1}$ must be in m different residue classes. By the definition of the complete system of residues, $y_0, \cdots, y_{m-1}$ is indeed a complete system of residues modulo m.

2. The proof is similar to that of step 1.

Theorem 2.11

1. *Let a, b be integers and $(a, m) = 1$, then when x runs over a complete system of residues modulo m, so does $ax + b$.*

2. *Let a, b be integers and $(a, m) = 1$, then when x runs over a reduced system of residues modulo m, so does $ax + bm$.*

Proof

1. Assume that $x_0, x_1, \cdots, x_{m-1}$ is a complete system of residues modulo m, then $x_0, x_1, \cdots, x_{m-1}$ are mutually incongruent modulo m. Since $(a, m) = 1$, for $0 \le i, j \le m - 1$, $ax_i + b \equiv ax_j + b \pmod{m}$ if and only if $x_i \equiv x_j \pmod{m}$. This implies that $ax_0 + b, \cdots, ax_{m-1} + b$ are mutually incongruent modulo m. Thus, the first part of the theorem is deduced from Theorem 2.10.

2. Assume that $x_0, x_1, \cdots, x_{\varphi(m)-1}$ is a reduced system of residues modulo m, then $x_0, x_1, \cdots, x_{\varphi(m)-1}$ are mutually incongruent modulo m. Since $(a, m) = 1$, we know that $ax_0 + bm, \cdots, ax_{\varphi(m)-1} + bm$ are mutually incongruent modulo m and coprime to m.

Thus, the second part of the theorem is deduced by Theorem 2.10.

Theorem 2.12 Let $m = m_1 m_2$, $(m_1, m_2) = 1$. If $x_i^{(1)} (0 \le i \le m_1 - 1)$ $(x_i^{(1)} (0 \le i \le \phi(m_1) - 1))$ run over a complete (reduced) system of residues modulo m_1, $x_j^{(2)}$ $(0 \le j \le m_2 - 1)$ $(x_j^{(2)}$ $(0 \le j \le \phi(m_2) - 1))$ run over a complete (reduced) system of residues modulo m_2, then $x_{ij} = m_2 x_i^{(1)} + m_1 x_j^{(2)}$ run over a complete (reduced) system of residues modulo m.

Proof We will first show that if $x_i^{(1)} (0 \le i \le m_1 - 1), x_j^{(2)} (0 \le j \le m_2 - 1)$ are complete systems residues modulo m_1, m_2, respectively, then $x_{ij} = m_2 x_i^{(1)} + m_1 x_j^{(2)}, 0 \le i < m_1, 0 \le j < m_2$ forms a complete system of residues modulo m.

Obviously, there are $m = m_1 m_2$ numbers of x_{ij}, so we only need to show that they are mutually incongruent. If

$$m_2 x_{i_1}^{(1)} + m_1 x_{j_1}^{(2)} \equiv x_{i_1 j_1} \equiv x_{i_2 j_2} \equiv m_2 x_{i_2}^{(1)} + m_1 x_{j_2}^{(2)} \pmod{m_1 m_2},$$

then

$$x_{i_1 j_1} \equiv x_{i_2 j_2} \pmod{m_1}, \quad x_{i_1 j_1} \equiv x_{i_2 j_2} \pmod{m_2}.$$

Therefore,

$$m_2 x_{i_1}^{(1)} \equiv m_2 x_{i_2}^{(1)} \pmod{m_1}, \quad m_1 x_{j_1}^{(2)} \equiv m_1 x_{j_2}^{(2)} \pmod{m_2}.$$

From the assumption $(m_1, m_2) = 1$,

$$x_{i_1}^{(1)} \equiv x_{i_2}^{(1)} \pmod{m_1}, x_{j_1}^{(2)} \equiv x_{j_2}^{(2)} \pmod{m_2}.$$

This proves that the $m_1 m_2$ numbers x_{ij} are mutually incongruent; hence, they form a complete system of residues modulo m.

Next, we show that if $x_i^{(1)}, x_j^{(2)}$ run over reduced systems of residues modulo m_1, m_2, respectively, then x_{ij} forms a reduced system of residues modulo m.

From the proof above, x_{ij} are mutually incongruent. So we only need to show that

$$(x_{ij}, m) = 1 \quad \text{if and only if} \quad (x_i^{(1)}, m_1) = (x_j^{(2)}, m_2) = 1.$$

Since

$$(m_1, m_2) = 1,$$

$(x_{ij}, m) = 1$ if and only if

$$(m_2 x_i^{(1)} + m_1 x_j^{(2)}, m_1) = 1, \quad (m_2 x_i^{(1)} + m_1 x_j^{(2)}, m_2) = 1,$$

if and only if

$$(x_i^{(1)}, m_1) = (x_j^{(2)}, m_2) = 1.$$

The theorem is proved.

It is easy to show that the conditions in Theorems 2.11 and 2.12 are necessary and sufficient conditions.

Theorem 2.13 *Let* $m = m_1 \cdots m_k$ *and* $m_1, \cdots, m_k$ *be mutually coprime. Let* $m = m_j M_j, 1 \leq j \leq k,$ *and*

$$x = M_1 x^{(1)} + \cdots + M_k x^{(k)}. \tag{2.2}$$

Then when x *runs over the complete (reduced) system of residues modulo* m *if and only if* $x^{(1)}, \cdots, x^{(k)}$ *run over the complete (reduced) system of residues modulo* $m_1, \cdots, m_k$*, respectively.*

Proof When $k = 2$, this is Theorem 2.12. Assume that for $k = n(n \geq 2)$, the result holds. When $k = n + 1$, $m = m_1 \cdots m_n m_{n+1}$, and x is given by (2.2):

$$\overline{x}^{(n)} = \frac{m}{m_1 m_{n+1}} x^{(1)} + \cdots + \frac{m}{m_n m_{n+1}} x^{(n)}.$$

By Theorem 2.12, we have

$$x = m_{n+1} \overline{x}^{(n)} + \frac{m}{m_{n+1}} x^{(n+1)}.$$

These give the desired result.

Now we describe a special method of constructing a complete system of residues and a reduced system of residues.

Theorem 2.14 *Let* $m = m_1 m_2$, $x_i^{(1)}$, $1 \le i \le m_1$, *be a complete system of residues modulo* m_1, *and* $x_j^{(2)}$, $1 \le j \le m_2$, *be a complete system of residues modulo* m_2, *then* $x_{ij} = x_i^{(1)} + m_1 x_j^{(2)}$ *is a complete system of residues modulo* m.

In general, if $m = m_1 \cdots m_k$, $x = x^{(1)} + m_1 x^{(2)} + \cdots + m_1 m_2 \cdots m_{k-1} x^{(k)}$, *then if* $x^{(1)}, \cdots, x^{(k)}$ *run over complete systems of residues modulo* $m_1, \cdots, m_k$, *respectively,* x *runs over a complete system of residues modulo* m.

Proof We first show that the result is true for $k = 2$. In this case, there are $m = m_1 m_2$ numbers of x_{ij}, and we only need to prove that they are mutually incongruent. If

$$x_{i_1}^{(1)} + m_1 x_{j_1}^{(2)} = x_{i_1 j_1} \equiv x_{i_2 j_2} = x_{i_2}^{(1)} + m_1 x_{j_2}^{(2)} \pmod{m_1 m_2},$$

then $x_{i_1}^{(1)} = x_{i_2}^{(1)} \pmod{m_1}$. Since the values $x_{i_1}^{(1)}, x_{i_2}^{(1)}$ are taken from the same residue class modulo m_1, we have $x_{i_1}^{(1)} = x_{i_2}^{(1)}$. From the above relation, we get $m_1 x_{j_1}^{(2)} \equiv m_1 x_{j_2}^{(2)} \pmod{m_1 m_2}$, that is, $x_{j_1}^{(2)} \equiv x_{j_2}^{(2)} \pmod{m_2}$, and a similar argument gives us $x_{j_1}^{(2)} = x_{j_2}^{(2)}$. Thus, the first part of the theorem is proved.

Assume that the result holds when $k = n (n \ge 2)$. We consider the case of $k = n + 1$. Now $m = m_1 \cdots m_n m_{n+1}$. Set

$$\bar{x}^{(n)} = x^{(1)} + m_1 x^{(2)} + \cdots + m_1 m_2 \cdots m_{n-1} x^{(n)},$$

and we get

$$x = \bar{x}^{(n)} + m_1 \cdots m_{n-1} m_n x^{(n+1)}.$$

By the proof of the first part, we get the desired result.

Remark Theorem 2.14 only gives a sufficient condition, not a necessary condition.

Theorems 2.11 through 2.13 show that a complete (reduced) system of residues for large modulus can be a combination of two complete (reduced) systems of residues of smaller modulus. In fact, construction of a complete (reduced) system of residues for large modulus can be done by factoring the modulus (by using primitive roots of prime powers through prime factorization).

Finally, we discuss the relationship between system residues for a modulus and systems of residues for factors of the modulus.

Theorem 2.15 *Let $m_1|m$. Then for any r, we have*

$$r \pmod{m} \subseteq r \pmod{m_1}.$$

The equality holds if and only if $m_1 = m$.

If $l_0, \cdots, l_{d-1}$ form a complete system of residues $d = m/m_1$, then the d congruence classes on the right-hand side of the following relation are pairwise distinct:

$$r \pmod{m_1} = \bigcup_{0 \le j \le d-1} (r + l_j m_1) \pmod{m}. \tag{2.3}$$

In particular,

$$r \pmod{m_1} = \bigcup_{0 \le j \le d-1} (r + j m_1) \pmod{m}.$$

Proof We partition the integers in the congruence class $r \pmod{m_1}$ according to the modulus m. For any two numbers $r + k_1 m_1, r + k_2 m_1$ in $r \pmod{m_1}$, the congruence

$$r + k_1 m_1 \equiv r + k_2 m_1 \pmod{m}$$

holds if and only if

$$k_1 \equiv k_1 \pmod{d}.$$

This implies that the d congruence classes modulo m on the right-hand side of (2.3) are mutually disjoint, and any number $r + k m_1$ of $r \bmod m_1$ must be in one of the congruence classes. On the other hand, for any j, we must have

$$(r + l_j m_1) \pmod{m} \subseteq (r + l_j m_1) \bmod m_1 = r \pmod{m_1}.$$

The theorem is proved.

Example 2.16 *For modulus $m = 5 \times 7$, construct a complete and reduced system of residues modulo m.*

Solution: Let $m_1 = 5, m_2 = 7, (5, 7) = 1$, then $M_1 = 7$, $M_2 = 5$, when $x^{(1)}, x^{(2)}$ run over a complete (reduced) system of residues modulo 5 and 7, respectively,

$$x = M_1 x^{(1)} + M_2 x^{(2)} = 7 x^{(1)} + 5 x^{(2)}$$

runs over a complete (reduced) system of residues modulo 35. The following table shows a complete system of residues modulo 35. If we exclude the underlined elements, the rest form a reduced system of residues modulo 35.

$x^{(1)}$	$x^{(2)}$						
	-3	-2	-1	0	1	2	3
-2	-29	-24	-19	$\underline{-14}$	-9	-4	1
-1	-22	-17	-12	$\underline{-7}$	-2	3	8
0	$\underline{-15}$	$\underline{-10}$	$\underline{-5}$	$\underline{0}$	$\underline{5}$	$\underline{10}$	$\underline{15}$
1	-8	-3	2	$\underline{7}$	12	17	22
2	-1	4	9	$\underline{14}$	19	24	29

Example 2.17 *Prove that if $n \geq 1$, then*

$$\sum_{d|n} \varphi(d) = n.$$

Proof We partition the set $S = \{1, 2, \cdots, n\}$ in the following manner:

$$S = \bigcup_d S_d, \quad S_d = \{m | (m, n) = d\}.$$

We have

$$m \in S_d \Leftrightarrow (m, n) = d \Leftrightarrow (m/d, n/d) = 1.$$

Notice that the number of integers of the form m/d that is coprime to n/d is $\varphi(n/d)$, and when d is fixed, the correspondence between m/d and m is one to one, so the number of elements in S_d is $\varphi(n/d)$. Therefore,

$$n = \sum_{d|n} \varphi(n/d) = \sum_{d|n} \varphi(d).$$

The proof is complete.

2.3 EULER'S THEOREM

Euler's function plays a very important role in number theory. Next, we will use congruence theory to give a property of Euler's function and an evaluation formula of Euler's function for integers with given prime factorizations.

Theorem 2.18

1. *Let p be prime, $k \geq 1$, then*

$$\varphi(p^k) = p^{k-1}(p-1).$$

2. *If $m = m_1 m_2$ and $(m_1, m_2) = 1$, then $\varphi(m) = \varphi(m_1)\varphi(m_2)$.*

3. *If $m = p_1^{\alpha_1} \cdots p_r^{\alpha_r}$, where $p_1, \cdots, p_r$ are distinct prime factors, then*

$$\varphi(m) = p_1^{\alpha_1 - 1}(p_1 - 1) \cdots p_r^{\alpha_r - 1}(p_r - 1) = m \prod_{p \mid m} \left(1 - \frac{1}{p}\right).$$

Proof

1. $\varphi(p^k)$ is the number of r's that satisfy the following condition: $(r, p^k) = 1, 1 \leq r \leq p^k$. Since p is a prime,

$$(r, p^k) = 1 \Leftrightarrow (r, p) = 1.$$

This means that $\varphi(p^k)$ is the number of r's in $1, 2, \cdots, p^k$ that are not divisible by p. However, the numbers in $1, 2, \cdots, p^k$ that are divisible by p are of the form of np, $n = 1, 2, \cdots, p^{k-1}$, so there are exactly p^{k-1} of them. Therefore, $\varphi(p^k) = p^k - p^{k-1} = p^{k-1}(p-1)$.

2. This is a consequence of Theorem 2.12 of Section 2.2.

3. From part 2, we can further conclude that if $m = m_1 m_2 \cdots m_r$ and $m_1, m_2, \cdots, m_r$ are mutually coprime, then

$$\varphi(m) = \varphi(m_1)\varphi(m_2 \cdots m_r) = \varphi(m_1)\varphi(m_2) \cdots \varphi(m_r).$$

In particular, if $m > 1$ and $m = p_1^{\alpha_1} \cdots p_r^{\alpha_r}$, then

$$\varphi(m) = p_1^{\alpha_1 - 1}(p_1 - 1) \cdots p_r^{\alpha_r - 1}(p_r - 1) = m \prod_{p \mid m} \left(1 - \frac{1}{p}\right).$$

The theorem is proved.

Remark From Theorem 2.18, except for $\varphi(1) = \varphi(2) = 1$, we have $2 \mid \varphi(m)$ for all $m \geq 3$.

Corollary 2.19 *The numbers $a + bp$, $(1 \leq a \leq p - 1, 0 \leq b \leq p^{k-1} - 1)$ forms a reduced system of residues modulo p^k.*

Proof By Theorem 2.14 of Section 2.2, we see that

$$r = a + bp, 1 \le a \le p-1, 0 \le b \le p^{k-1} - 1$$

runs over a reduced system of residues modulo p^k.

A reduced system of residues modulo m can take various different forms; however, it is not difficult to see that the product of all numbers in the system is an invariant modulo m. That is, if $\{r_0, \cdots, r_{\varphi(m)-1}\}, \{r'_0, \cdots, r'_{\varphi(m)-1}\}$ are two reduced systems of residues modulo m, then we must have

$$\prod_{j=1}^{\varphi(m)} r_j \equiv \prod_{j=1}^{\varphi(m)} r'_j \pmod{m}.$$

This can be used to get the following famous Euler's theorem.

Theorem 2.20 (*Euler's theorem*) *Let* $(a, m) = 1$, *then*

$$a^{\varphi(m)} \equiv 1 \pmod{m}.$$

Proof Let $r_0, \cdots, r_{\varphi(m)-1}$ be a reduced system of residues modulo m, by Theorem 2.11, if $(a, m) = 1$, $ar_0, \cdots, ar_{\varphi(m)-1}$ is also a reduced system of residues modulo m; thus,

$$\prod_{j=0}^{\varphi(m)-1} r_j \equiv \prod_{j=0}^{\varphi(m)-1} (ar_j) \equiv a^{\varphi(m)} \prod_{j=0}^{\varphi(m)-1} r_j \pmod{m}.$$

Since $(r_j, m) = 1, j = 0, \cdots, \varphi(m) - 1$,

$$a^{\varphi(m)} \equiv 1 \pmod{m}.$$

We have proved the theorem.

In particular, when p is a prime, $\varphi(p) = p - 1$. Thus, for any a with $(a, p) = 1$, we have

$$a^{p-1} \equiv 1 \pmod{p}.$$

This relation is usually called Fermat's little theorem.

Remark

1. In Euler's theorem, if $a = -1$, then we have $(-1)^{\varphi(m)} - 1 \equiv 0$ (mod m). This also shows that $2 | \varphi(m)$ if $m \geq 3$.

2. Euler's theorem gives a convenient method of finding the inverse a^{-1} of a modulo m, that is, when $(a, m) = 1$,

$$a^{-1} \equiv a^{\varphi(m)-1} \quad (\text{mod } m).$$

However, from the computational complexity point of view, this is not efficient. To use the above formula to compute a^{-1}, we need to compute $\varphi(m)$ first. So far, we have not had an efficient algorithm to evaluate $\varphi(m)$ for a general m.

Example 2.21 *Compute the last digit of the decimal representation of 7^{10001}.*

Solution: What we need to find is 7^{10001} (mod 10). Since

$$\varphi(10) = \varphi(2) \cdot \varphi(5) = 4,$$

and $(7, 10001) = 1$, by Euler's theorem we have

$$7^{10001} = 7^{4*2500+1} \equiv 7 \quad (\text{mod } 10).$$

That is, the last digit of the decimal representation of 7^{10001} is 7.

2.4   WILSON'S THEOREM

In this section, we introduce an important theorem regarding the product of numbers in a reduced system modulo m—Wilson's theorem.

Theorem 2.22 (*Wilson's theorem*) *Let p be a prime, $r_1, \cdots, r_{p-1}$ is a reduced system of residues modulo p, then*

$$\prod_{r \bmod p} r \equiv r_1 \cdots r_{p-1} \equiv -1 \quad (\text{mod } p).$$

In particular,

$$(p-1)! \equiv -1 \quad (\text{mod } p).$$

Proof The result is true when $p = 2$. Now assume that $p \geq 3$. For each r_i, $0 \leq i < p$, there is one and only one r_j such that

$$r_i r_j \equiv 1 \pmod{p}. \tag{2.4}$$

If $r_i r_j \equiv 1 \pmod{p}$, then

$$r_i = r_j \Leftrightarrow r_i^2 \equiv 1 \pmod{p}.$$

This means that

$$(r_i - 1)(r_i + 1) \equiv 0 \pmod{p}.$$

Since p is a prime and $p \geq 3$, this is equivalent to

$$r_i - 1 \equiv 0 \pmod{p} \quad \text{or} \quad r_i + 1 \equiv 0 \pmod{p}.$$

However, these two relations cannot hold simultaneously. This implies that in this reduced system of residues, for each r_i, except for $r_i \equiv 1, -1 \pmod{p}$, we must have $r_i \neq r_j$ that both make (2.4) hold true. We may assume that $r_1 \equiv 1 \pmod{p}, r_{p-1} \equiv -1 \pmod{p}$. Thus, except for r_1 and r_{p-1}, the rest of the numbers in the reduced system of residues are partitioned in pairs according to the relation in (2.4), so

$$r_2 \cdots r_{p-2} \equiv 1 \pmod{p}.$$

This proves the first part of the theorem.

Notice that $1, 2, \cdots, p-1$ is a reduced system of residues modulo p, so the second part of the theorem follows.

A similar result holds for modulus p^l.

Theorem 2.23 *Let $p \geq 3$ be a prime, and $c = \varphi(p^l)$ with $l \geq 1$. Let $r_1, r_2, \cdots, r_c$ be a reduced system of residues modulo p^l. We have*

$$r_1 r_2 \cdots r_c \equiv -1 \pmod{p^l}.$$

In particular,

$$\prod_{r=1}^{p-1} \prod_{s=0}^{p^{l-1}-1} (r + ps) \equiv -1 \pmod{p^l}.$$

Following the notation and condition of Theorem 2.18, we have

$$c = \varphi(p^l) = \varphi(2p^l).$$

Set

$$r'_j = \begin{cases} r_j, & \text{if } r_j \text{ is odd}, \\ r_j + p^l, & \text{if } 2|r_j. \end{cases}$$

Then it is easy to see that $r'_j(1 \le j \le c)$ is still a reduced system of residues modulo p^l, and all of its members are odd numbers. This means that they are also a reduced system of residues modulo $2p^l$. Furthermore,

$$r'_1 \cdots r'_c \equiv -1 \pmod{2p^l}.$$

This implies the following result.

Theorem 2.24 *Let $p \ge 3$ be a prime and $c = \varphi(2p^l)$ with $l \ge 1$. Let $r_1, r_2, \cdots, r_c$ be a reduced system of residues modulo $2p^l$. Then we have*

$$r_1 r_2 \cdots r_c \equiv -1 \pmod{2p^l}.$$

Theorem 2.25 *Let $c = \varphi(2^l) = 2^{l-1}$ with $l \ge 1$, and $r_1, \cdots, r_c$ be a reduced system of residues modulo 2^l. We have*

$$r_1 \cdots r_c = \begin{cases} -1 \pmod{2^l}, & l = 1, 2, \\ 1 \pmod{2^l}, & l \ge 3. \end{cases}$$

Proof When $l = 1, 2$, the result can be verified directly. Now let us assume $l \ge 3$. For each r_i, there must be a unique r_j such that

$$r_i r_j \equiv 1 \pmod{2^l}. \tag{2.5}$$

It is noted that the pair r_i, r_j in (2.5) satisfies $r_i = r_j$ if and only if $r_i^2 \equiv 1 \pmod{2^l}$; that is,

$$(r_i - 1)(r_i + 1) \equiv 0 \pmod{2^l}.$$

As $(r_i, 2) = 1$, so we have

$$\frac{r_i - 1}{2} \cdot \frac{r_i + 1}{2} \equiv 0 \pmod{2^{l-2}}.$$

The fact that

$$\left(\frac{r_i - 1}{2}, \frac{r_i + 1}{2} \right) = 1$$

implies that $r_i = r_j$ if and only if

$$\frac{r_i - 1}{2} \equiv 0 \pmod{2^{l-2}} \quad \text{or} \quad \frac{r_i + 1}{2} \equiv 0 \pmod{2^{l-2}},$$

that is, if and only if

$$r_i \equiv 1 \pmod{2^{l-1}} \text{ or } r_i \equiv -1 \pmod{2^{l-1}}.$$

This shows that only if

$$r_i \equiv 1, 2^{l-1} + 1, 2^{l-1} - 1, \text{ or } 2^l - 1 \pmod{2^l},$$

we have $r_i = r_j$ for the pair r_i, r_j in (2.5). So, for each r_i in this reduced system of residues other than the above for numbers (which are mutual incongruent modulo 2^l), there must be $r_j \neq r_i$ such that pair r_i, r_j satisfies (2.5). These $c - 4$ numbers in the reduced system residues are partitioned in pairs according to (2.5), and hence their product is congruent to 1 modulo 2^l. This, together with the previous relations, proves the theorem.

In summary, we have proved when $m = 2, 4, p^l, 2p^l$ (p is an odd prime), the product of numbers in a reduced system of residues modulo m is congruent to -1 modulo m. It can be proved that for other modulus m, such product must be congruent to 1 modulo m.

Wilson's theorem can simplify some operations in certain cases.

Example 2.26 *Let p be a prime and $1 \leq n \leq p - 1$. Prove that*

$$(-1)^n n!(p - n - 1)! \equiv -1 \pmod{p}.$$

Proof By Wilson's theorem,

$$(p - 1)(p - 2) \cdots (p - n)(p - n - 1)! \equiv -1 \pmod{p}.$$

After expansion, this yields

$$[p^n - (1 + 2 + \cdots + n)p^{n-1} + \cdots + (-1)^n n!](p - n - 1)! \equiv -1 \pmod{p}.$$

From the property of congruence, we have

$$(-1)^n n!(p - n - 1)! \equiv -1 \pmod{p}.$$

The proof is complete.

EXERCISES

2.1 Find

  (1) 2^{143} (mod 13).

  (2) 5^{141} (mod 47).

2.2 Determine whether the following are true, and explain why.

  (1) If $a^3 \equiv b^3$ (mod m), then $a \equiv b$ (mod m).

  (2) If $a \equiv b$ (mod m), then $a^3 \equiv b^3$ (mod m).

  (3) If $ac \equiv bc$ (mod m), then $a \equiv b$ (mod m).

  (4) If $a_1 \equiv a_2$ (mod m) and $b_1 \equiv b_2$ (mod m), then $(a_1)^{b_1} \equiv (a_2)^{b_2}$ (mod m).

2.3 Find the inverses of all elements modulo $m = 7, 13$, respectively.

2.4 Prove that for any positive integer n, at least one of the following congruences holds:

$$n \equiv 0 \ (\text{mod } 2), \quad n \equiv 0 \ (\text{mod } 3), \quad n \equiv 1 \ (\text{mod } 4),$$

$$n \equiv 3 \ (\text{mod } 18), \quad n \equiv 7 \ (\text{mod } 12), \quad n \equiv 23 \ (\text{mod } 24).$$

2.5 Prove that if $m > 2$, then $0^2, 1^2, \cdots, (m-1)^2$ cannot be a complete system of residues modulo m.

2.6 Let n, h be positive integers. Prove that among the positive integers that are not larger than nh, there are $h\varphi(n)$ numbers that are coprime to n.

2.7 Let $m = m_1 m_2 \cdots m_k$, $m_1, m_2, \cdots, m_k$ be mutually coprime, and $(m, a_i) = 1$. Prove that if $x^{(i)}$ runs over a complete (reduced) system of residues modulo m_i for each i,

$$x = (M_1 a_1 x^{(1)} + M_2 + \cdots + M_k)(M_1 + M_2 a_2 x^{(2)} + \cdots + M_k)$$
$$\cdots (M_1 + M_2 + \cdots + M_k a_k x^{(k)})$$

runs over a reduced system of residues modulo $m = m_1 m_2 \cdots m_k$, where $m = M_i m_i, 1 \leq i \leq k$.

2.8 For modulus $m = 7 \times 11$, construct a reduced system of residues and a complete system of residues modulo m.

2.9 Let n be an integer, find a sufficient and necessary condition n and divisible by 11, in terms of the decimal digits of n.

2.10 Let p be an odd prime and prove that

(1) If $p = 4m + 3$, then for any integer a we have $a^2 \not\equiv -1$ (mod p).

(2) If $p = 4m + 1$, then there exists an integer a such that $a^2 \equiv -1$ (mod p).

2.11 Let m, n be coprime integers, and prove that

$$m^{\varphi(n)} + n^{\varphi(m)} \equiv 1 \pmod{mn}.$$

2.12 Let p be an odd prime, and prove that

$$(p-1)! \equiv p - 1 \left(\bmod \frac{p(p-1)}{2} \right).$$

2.13 Prove that there are no $x, y \in \mathbb{Z}$ such that the equation $y^2 = x^3 + 7$ holds.

CHAPTER 3

Congruence Equations

T HE MAIN focus of this chapter is to introduce the solvability of congruence equations in one variable and systems of linear congruence equations and provide some concrete methods for finding their solutions. For a system of linear congruence equations, we will mainly discuss how to solve them using the Chinese remainder theorem. For general congruence equations, we describe a general process of finding solutions. For quadratic congruence equations, we consider the case of prime modulus, that is, the quadratic residue problem with prime modulus. Finally, we introduce an arithmetic function that is related to the quadratic residues, the Legendre symbol, and define a more general arithmetic function, the Jacobi symbol.

The content of this chapter consists of not only key materials for the congruence theory in elementary number theory, but also the most important and fundamental theory for public key cryptography. Various methods of solving congruence equations can contribute significantly to the design and analysis of many cryptographic algorithms. For example, solving linear congruence equations is one of the most basic operations for the encryption, decryption, and even breaking of many cryptosystems; the Chinese remainder theorem has played important roles in efficient implementations of several cryptographic systems, and it also has a direct application to the design of cryptographic systems of special forms; quadratic residues and the Jacobi symbol can be used in the primality testing and pseudorandom number generators.

3.1 BASIC CONCEPTS OF CONGRUENCES OF HIGH DEGREES

In this section, we will mainly introduce general congruence equations and related concepts. We will also touch on simplified forms of congruence equations. Unless otherwise stated, the congruence equation mentioned here will be a congruence equation with one variable.

Definition 3.1 Let $f(x)$ be a polynomial with integer coefficients

$$f(x) = a_n x^n + \cdots + a_1 x + a_0,$$

then the congruence involving a variable

$$f(x) \equiv 0 \pmod{m} \tag{3.1}$$

is called a congruence equation modulo m. If an integer c satisfies

$$f(c) \equiv 0 \pmod{m},$$

then c is called a solution to the congruence equation $f(x) \equiv 0 \pmod{m}$. □

It is obvious that if c is a solution to the congruence equation $f(x) \equiv 0 \pmod{m}$, then any integer in the congruence class $c \pmod{m}$ is also a solution of the equation, so we say that the congruence class $c \pmod{m}$ is a solution of $f(x) \equiv 0 \pmod{m}$. The number of all pairwise noncongruent solutions modulo m is called the number of solutions of the congruence equation of $f(x) \equiv 0 \pmod{m}$.

Definition 3.2 If $m \nmid a_n$, then the degree of the congruence equation (3.1) is n; if $m \mid a_j, k+1 \leq j \leq n$ and $m \nmid a_k$, then the degree of the congruence equation is k. □

From Definition 3.2, the degree of the congruence equation (3.1) is not necessarily the degree of the polynomial $f(x)$. Obviously, the number of solutions of a congruence equation modulo m is at most m, and we can solve a congruence equation by validating the complete residue system modulo m. We can also find a solution by simplifying the congruence equation via identical transformation. Several major identical transformations are as follows.

Property 1 If $f(x) \equiv g(x) \pmod{m}$, then the congruence equations $f(x) \equiv 0 \pmod{m}$ and $g(x) \equiv 0 \pmod{m}$ have the same set (hence, the same number) of solutions.

Property 2 If $f(x) = q(x)h(x) + r(x)$ and the congruence equation $h(x) \equiv 0 \pmod{m}$ is actually an identity, that is, the number of solutions of the equation is m, then the congruence equation (3.1) and the congruence equation

$$r(x) \equiv 0 \pmod{m}$$

have the same set (hence, the same number) of solutions.

The key step of using identities to reduce the degree of congruence equations is to find identities modulo m. If m is a prime p, then from the Fermat's little theorem, we know that

$$h(x) = x^p - x \equiv 0 \pmod{p}$$

is an identity.

Property 3 If $(a, m) = 1$, then the congruence equations

$$f(x) \equiv 0 \pmod{m}$$

and

$$af(x) \equiv 0 \pmod{m}$$

are equivalent.

In particular, if $(a_n, m) = 1$, then the congruence equation (3.1) can be turned to a congruence equation with leading coefficient 1:

$$(a_n)^{-1} f(x) \equiv 0 \pmod{m}.$$

Using properties of congruences, we can get a necessary condition for a congruence equation having the solution.

Theorem 3.3 *If integer $d \mid m$, then a necessary condition for the congruence equation $f(x) \equiv 0 \pmod{m}$ having a solution is*

$$f(x) \equiv 0 \pmod{d}$$

has a solution.

This theorem can be used to determine when a congruence equation does not have a solution.

Example 3.4 *Solve the congruence equation $f(x) = 4x^2 - 27x - 9 \equiv 0$ (mod 15).*

Solution: The prime factors of 15 are 3 and 5. However, we only have $f(3) \equiv 0 \pmod 3$. Since $f(1), f(2), f(3), f(4), f(5) \not\equiv 0 \pmod 5$, this equation has no solution.

Example 3.5 *Solve the congruence equation* $5x^3 - 3x^2 + 3x - 1 \equiv 0$ (mod 11).

Solution: Consider the complete residues of 11 in least absolute values: $-5, -4, -3, -2, -1, 0, 1, 2, 3, 4, 5$. A direct calculation shows that $x = 2$ is a solution. So the solution of this congruence equation is $x \equiv 2 \pmod{11}$.

Example 3.6 *Solve the congruence equation*

$$3x^{15} - x^{13} - x^{12} + x^{11} - 3x^5 + 6x^3 - 2x^2 + 2x - 1 \equiv 0 \pmod{11}.$$

Solution: Using the identity $x^{11} \equiv x \pmod{11}$ and the polynomial division

$$3x^{15} - x^{13} - x^{12} + x^{11} - 3x^5 + 6x^3 - 2x^2 + 2x - 1$$
$$= (x^{11} - x)(3x^4 - x^2 - x + 1) + 5x^3 - 3x^2 + 3x - 1 \pmod{11},$$

we know that the original equation and the congruence equation $5x^3 - 3x^2 + 3x - 1 \equiv 0 \pmod{11}$ have the same solutions. From Example 3.5, we see that $x \equiv 2 \pmod{11}$ is the solution of the congruence equation.

3.2 LINEAR CONGRUENCES

We discussed general congruence equations in Section 3.1. The main topic for this section is the solvability and methods of solving the linear congruence equation of the following form:

$$ax \equiv b \pmod{m}. \tag{3.2}$$

Theorem 3.7 *If $(a, m) = 1$, then the congruence equation $ax \equiv b$ (mod m) has one and only one solution.*

Proof If $(a, m) = 1$, then there is an a^{-1} such that $aa^{-1} \equiv 1$ (mod m). Therefore, $x \equiv a^{-1}b \pmod{m}$ satisfies the equation. Assume that $x'(\bmod m)$ is another solution of the equation, then $ax \equiv ax'$ (mod m). Since $(a, m) = 1$, so $x \equiv x' \pmod{m}$. The proof is complete.

From Euler's theorem, if $(a, m) = 1$, $a^{-1} = a^{\phi(m)-1}$ (mod m), so the unique solution of the congruence equation (3.2) is $x = a^{\phi(m)-1}b$ (mod m). In practice, an efficient method of solving equation (3.2) is to find a^{-1} using the Euclidean algorithm.

Theorem 3.8 *Congruence equation (3.2) has a solution if and only if $(a, m)|b$. If this equation has a solution, then the number of its solutions is (a, m). If x_0 is a fixed solution, then all of the (a, m) solutions are*

$$x \equiv x_0 + \frac{m}{(a, m)}t \quad (\text{mod } m), \ t = 0, 1, \cdots, (a, m) - 1.$$

Proof Necessity: If the linear congruence equation (3.2) has a solution, then there are x_1, y_1 such that $ax_1 = b + my_1$, so $(a, m)|b$.

Sufficiency: Let $d = (a, m)$. If $(a, m) \mid b$, then by property 5 of Chapter 2, congruence $ax \equiv b$ (mod m) holds if and only if $\frac{a}{d}x \equiv \frac{b}{d}$ (mod $\frac{m}{d}$) holds. Since $(\frac{a}{d}, \frac{m}{d}) = 1$,

$$\frac{a}{d}x \equiv \frac{b}{d} \left(\text{mod } \frac{m}{d} \right) \tag{3.3}$$

has a solution, and hence, $ax \equiv b$ (mod m) has a solution. The sufficiency is proved.

If x_0 is a fixed solution of equation (3.2) and x is an arbitrary solution of (3.2), then both x_0 (mod $\frac{m}{d}$) and x (mod $\frac{m}{d}$) are the unique solution of (3.3), so $x \equiv x_0$ (mod $\frac{m}{d}$). Conversely, for any $x \in \mathbb{Z}$, if $x \equiv x_0$ (mod $\frac{m}{d}$), then x is a solution of (3.3). Therefore, all solutions of the congruence equation (3.2) are

$$x_0 + \frac{m}{d}t \text{ mod } m, \ t = 0 \cdots, d - 1.$$

The theorem is proved.

The discussion above gives the solvability and number of solutions of the equation $ax \equiv b$ (mod m). What we describe next are the steps of finding solutions of $ax \equiv b$ (mod m):

1. Use identity transformation to transform the equation to $a'x \equiv b'$ (mod m), where $-m/2 < a' \leq m/2, -m/2 < b' \leq m/2$.

2. The congruence equation $a'x \equiv b'$ (mod m) and the Diophantine equation $a'x = my + b'$ either have a solution or have no solution simultaneously. Hence, $a'x \equiv b'$ (mod m) and $my \equiv -b'$ (mod $|a'|$) either have a solution or have no solution simultaneously.

3. If a solution of $my \equiv -b'$ (mod $|a'|$) is y_0 (mod $|a'|$), then $x_0 = (my_0 + b')/a'$ (mod m) is a solution of $a'x \equiv b'$ (mod m).

In fact, these steps are essentially the Euclidean algorithm for least absolute residues. It is the same as finding a special solution of a linear congruence in one variable. We will demonstrate this using a concrete example.

Example 3.9 *Solve congruence equation* $11x \equiv 217$ (mod 1732).

Solution: The equation has a solution if and only if $5y \equiv 3$ (mod 11) has a solution. Similarly,

$$5y \equiv 3 \quad (\text{mod } 11) \Leftrightarrow u \equiv 2 \quad (\text{mod } 5).$$

Therefore,

$$y \equiv (11 \times 2 + 3)/5 \quad (\text{mod } 11) \equiv 5 \quad (\text{mod } 11),$$
$$x \equiv (1732 \times 5 + 217)/11 \quad (\text{mod } 1732) \equiv 807 \quad (\text{mod } 1732).$$

3.3 SYSTEMS OF LINEAR CONGRUENCE EQUATIONS AND THE CHINESE REMAINDER THEOREM

Definition 3.10 Let $f_i(x)$ be a polynomial with integer coefficients, $1 \leq i \leq k$, we call the system of congruence relations with variable x

$$f_i(x) \equiv 0 \quad (\text{mod } m_i), \ 1 \leq i \leq k, \tag{3.4}$$

a system of congruence equations. If an integer c satisfies

$$f_i(c) \equiv 0 \quad (\text{mod } m_i), \ 1 \leq i \leq k,$$

simultaneously, then c is called a solution of the system of congruence equations. □

If c is a solution of the system of congruence equations (3.4), then any integer in the congruence class $c \bmod m, m = [m_1, \cdots, m_k]$ is also a solution of congruence equations (3.4). Therefore, c (mod m) can be regarded as a solution of the system of congruence equations. The number of all mutually incongruent solutions modulo m of the system of congruence equations is called the number of solutions of the system of congruence equations.

Obviously, the system of congruence equations (3.4) has at most m solutions, and it has no solution if one of the equations does not have solution. Now we discuss the solvability of a system of linear congruence equations when $m_0, \cdots, m_{k-1}$ are mutually coprime.

Theorem 3.11 *(The Chinese remainder theorem) Let $m_0, \cdots, m_{k-1}$ be positive integers that are pairwise coprime, then for any integers $a_0, \cdots, a_{k-1}$, the system of linear congruence equations*

$$x \equiv a_i \pmod{m_i}, \ 0 \le i \le k-1, \tag{3.5}$$

has a unique solution. This solution is

$$x \equiv M_0 M_0^{-1} a_0 + \cdots + M_{k-1} M_{k-1}^{-1} a_{k-1} \pmod{m},$$

where

$$m = m_0 \cdots m_{k-1} = m_i M_i, \ (0 \le i \le k-1)$$

and

$$M_i M_i^{-1} \equiv 1 \pmod{m_i}, \ (0 \le i \le k-1).$$

Proof First, let us prove $x \equiv M_0 M_0^{-1} a_1 + \cdots + M_{k-1} M_{k-1}^{-1} a_{k-1}$ (mod m) is indeed a solution of equations (3.5). By the fact that $M_i M_i^{-1} \equiv 1 \pmod{m_i}$, $m_i | M_j$, $i \ne j$, we have

$$x \equiv M_i M_i^{-1} a_i \equiv a_i \pmod{m_i}, \ 0 \le i \le k-1;$$

therefore, x is a solution of equations (3.5).

Next, we prove the uniqueness. If there are two solutions x_1, x_2, then we must have

$$x_1 \equiv x_2 \equiv a_i \pmod{m_i}, \ 0 \le i \le k-1.$$

Thus,

$$m_i | x_1 - x_2, \ 0 \le i \le k-1.$$

Since $m_1, m_2, \cdots, m_{k-1}$ are mutually coprime, $m = [m_0, \cdots, m_{k-1}] = m_0 \cdots m_{k-1}$ holds. Therefore, $x_1 \equiv x_2 \pmod{m}$, and the theorem is proved.

The Chinese remainder theorem describes a method of finding the solution of a system of linear congruence equations under the assumption that $m_0, \cdots, m_{k-1}$ are pairwise coprime. In this case, we say that the system of linear congruence equations (3.5) satisfies the condition of the Chinese remainder theorem. For a more general set of modulus $m_0, \cdots, m_{k-1}$, we can transform the system of congruence equations into the one that satisfies the condition of the Chinese remainder theorem through prime factorization or finds the greatest common divisors to factorize $m_0, \cdots, m_{k-1}$.

Example 3.12 *Solve the system of congruence equations*

$$\begin{cases} x \equiv 2 \pmod 3, \\ x \equiv 2 \pmod 5, \\ x \equiv -3 \pmod 7, \\ x \equiv -2 \pmod{13}. \end{cases}$$

Solution: By the Chinese remainder theorem,

$$M_1 = 455, \quad M_2 = 273, \quad M_3 = 195, \quad M_4 = 105;$$
$$M_1^{-1} = -1, \quad M_2^{-1} = 2, \quad M_2^{-1} = -1, \quad M_4^{-1} = 1.$$
$$\begin{aligned} x &= 455 \times (-1) \times 2 + 273 \times 2 \times 2 + 195 \times (-1) \times (-3) + 105 \times 1 \times (-2) \\ &= (-910) + 1092 + 585 + (-210) \\ &= 557 \pmod{1365}. \end{aligned}$$

Example 3.13 *Solve the system of congruence equations*

$$\begin{cases} 4x \equiv 14 \pmod{15}, \\ 9x \equiv 11 \pmod{20}. \end{cases}$$

Solution: We cannot use the Chinese remainder theorem directly since in this system of congruence equations, 20 and 15 are not coprime. This system of congruence equations is equivalent to the following:

$$\begin{cases} 4x \equiv 14 \pmod 5, \\ 4x \equiv 14 \pmod 3, \\ 9x \equiv 11 \pmod 4, \\ 9x \equiv 11 \pmod 5. \end{cases}$$

After some simplifications, we get

$$\begin{cases} x \equiv 1 \pmod 5, \\ x \equiv 2 \pmod 3, \\ x \equiv -1 \pmod 4, \\ x \equiv -1 \pmod 5. \end{cases}$$

The first equation and the fourth equation are conflicting, so the system of congruence equations has no solution.

3.4 GENERAL CONGRUENCE EQUATIONS

We have complete answers to the problems of solving a linear congruence equation and a system of linear congruence equations in the previous sections. However, there are no general methods for solving congruence equations of higher degrees. Next, we will give a brief and theoretical description about the main steps of solving general congruence equations.

Theorem 3.14 *If $m = m_0 m_1 \cdots m_{k-1}$, and m_i $(0 \leq i \leq k-1)$ are pairwise coprime, then the congruence equation*

$$f(x) \equiv 0 \pmod{m}$$

and the system of congruence equations

$$f(x) \equiv 0 \pmod{m_i}, \ 0 \leq i \leq k-1$$

have the same solution.

The proof can be obtained by using properties of congruence; we leave the detailed proof to the reader.

From Theorem 3.14, we know that in order to solve the congruence equation (3.1), it suffices to solve the system of congruence equations (3.4). To solve the system of congruence equations (3.4), we need to get solutions $a_{i1}, a_{i2}, \cdots, a_{il}$ of each of the congruence equations $f(x) \equiv 0 \pmod{m_i}$. For each j with $1 \leq j \leq l$, we solve the system of congruence equations $x \equiv a_{ij} \pmod{m_i}$, $0 \leq i \leq k-1$ and thus get a solution to $f(x) \equiv 0 \pmod{m}$.

Assuming $m = p_0^{\alpha_0} p_1^{\alpha_1} \cdots p_{k-1}^{\alpha_{k-1}}$, then the problem is reduced to finding solutions of the congruence equation

$$f(x) \equiv 0 \pmod{p_i^{\alpha_i}}.$$

Theorem 3.15 *If $x \equiv c_1, c_2, \cdots, c_s \pmod{p^{\alpha-1}}$ are solutions of $f(x) \equiv 0 \pmod{p^{\alpha-1}}$, then the solution of the equation*

$$f(x) \equiv 0 \pmod{p^{\alpha}} \tag{3.6}$$

that satisfies

$$x \equiv c_j \pmod{p^{\alpha-1}}, \ (1 \leq j \leq s)$$

has the form of $x \equiv c_j + p^{\alpha-1} y \pmod{p^{\alpha}}$, where y is a solution of

$$f'(c_j) y \equiv -f(c_j) p^{1-\alpha} \pmod{p}.$$

Proof Suppose that we have the solutions $x \equiv c_1, c_2, \cdots, c_s$ (mod $p^{\alpha-1}$) of $f(x) \equiv 0$ (mod $p^{\alpha-1}$), then for each solution of $f(x) \equiv 0$ (mod p^α), there is one and only one c_j ($1 \leq j \leq s$) such that $\alpha \equiv c_j$ (mod $p^{\alpha-1}$). Therefore, a solution of $f(x) \equiv 0$ (mod p^α) is of the form $x = c_j + p^{\alpha-1}y$. Notice that

$$
\begin{aligned}
f(c_j + p^{\alpha-1}y) &= a_n(c_j + p^{\alpha-1}y)^n + a_{n-1}(c_j + p^{\alpha-1}y)^{n-1} + \cdots \\
&\quad + a_1(c_j + p^{\alpha-1}y) + a_0 \\
&\equiv f(c_j) + p^{\alpha-1}f'(c_j)y + A_2 p^{2(\alpha-1)}y^2 + \cdots + A_n p^{n(\alpha-1)}y^n \\
&\equiv f(c_j) + p^{\alpha-1}f'(c_j)y \pmod{p^\alpha}.
\end{aligned}
$$

This implies that $f'(c_j)y \equiv -f(c_j)p^{1-\alpha}$ (mod p).

In summary, if we can solve the congruence equation $f(x) \equiv 0$ (mod p) modulo p, then we can solve the congruence equations $f(x) \equiv 0$ (mod p^i), $2 \leq i \leq \alpha$ for modulus $p^2, \cdots, p^\alpha$, and finally solve $f(x) \equiv 0$ (mod m).

Example 3.16 *Solve the congruence equation $x^2 \equiv 3$ (mod 11^3).*

Solution: A complete residue system modulo 11 can be represented as

$$
x = x_0 + x_1 \cdot 11 + x_2 \cdot 11^2, \quad -5 \leq x_i \leq 5, \ 0 \leq i \leq 2.
$$

We solve the following congruence equations in order:

$$
(x_0 + \cdots + x_i \cdot 11^i)^2 \equiv 3 \pmod{11^{i+1}}, \ 0 \leq i \leq 2.
$$

When $i = 0$, we solve $x_0^2 \equiv 3$ (mod 11) and get

$$
x_0 \equiv \pm 5 \pmod{11}.
$$

When $i = 1$, we solve $(\pm 5 + 11x_1)^2 \equiv 3$ (mod 11^2) and get

$$
\begin{aligned}
25 \pm 2 \times 5 \times 11 x_1 &\equiv 3 \pmod{11^2}, \\
x_1 &\equiv \pm 2 \pmod{11}.
\end{aligned}
$$

When $i = 2$, we solve $(\pm 5 \pm 2 \times 11 + 11^2 x_2)^2 \equiv 3$ (mod 11^3) and get

$$
\begin{aligned}
\pm 54 x_2 &\equiv -6 \pmod{11}, \\
x_2 &\equiv \mp 5 \pmod{11}.
\end{aligned}
$$

So,

$$
x = \pm 5 \pm 2 \times 11 \mp 5 \times 11^2 \equiv \pm 578 \pmod{11^3}.
$$

Example 3.17 *Solve the congruence equation $x^2 \equiv 1 \pmod{2^l}$.*

Solution: When $l = 1$, there is only one solution: $x \equiv 1 \pmod 2$. When $l = 2$, there are: $x \equiv -1, 1 \pmod{2^2}$. When $l \geq 3$, the congruence equation can be written as

$$(x-1)(x+1) \equiv 0 \pmod{2^l}.$$

When x is a solution, then it can be represented as $x = 2y + 1$. By plugging it into the above relation we get

$$4y(y+1) \equiv 0 \pmod{2^l},$$

that is,

$$y(y+1) \equiv 0 \pmod{2^{l-2}},$$

so we have

$$y \equiv 0, -1 \pmod{2^{l-2}}.$$

Therefore, the solution x must satisfy

$$x \equiv 1, -1 \pmod{2^{l-1}}.$$

This means that there are four solutions of the equation: $x \equiv 1$, $1 + 2^{l-1}, -1, -1 + 2^{l-1} \pmod{2^l}$.

3.5 QUADRATIC RESIDUES

In the discussion of Section 3.4, solving a general congruence equation is finally reduced to solving a congruence equation modulo prime number. One of the most common quadratic congruence equations modulo m is the class of quadratic residues, especially for the cases of $m = pq$ or $m = p$. In general, this type of congruence equation is transformed to the quadratic residue problem. The quadratic residue problem has extremely important applications in cryptography. In this section, we discuss quadratic congruence equations modulo odd primes.

The general form of a quadratic congruence equation is

$$ax^2 + bx + c \equiv 0 \pmod p.$$

If $(p, a) = 1$, then the above congruence equation can be simplified to

$$x^2 \equiv d \pmod p. \tag{3.7}$$

Notice that when $p|d$, $x^2 \equiv d \pmod p$ has one and only one solution $x \equiv 0 \pmod p$, so we will always assume $(p, d) = 1$ in the rest of our discussion.

Definition 3.18 Let prime $p > 2$, $(p, d) = 1$. If the congruence equation $x^2 \equiv d \pmod{p}$ has a solution, then d is said to be a quadratic residue modulo p; otherwise, d is said to be a quadratic nonresidue modulo p. We denote the sets of quadratic residues and nonresidues as

$$QR_p = \{a | a \in \mathbb{Z}_p^*, \text{ there exists } x \in \mathbb{Z}_p^*, x^2 \equiv a \pmod{p}\},$$
$$QNR_p = \{a | a \in \mathbb{Z}_p^*, \text{ for any } x \in \mathbb{Z}_p^*, x^2 \not\equiv a \pmod{p}\},$$

respectively. □

Theorem 3.19 *In the system of reduced residues modulo p, quadratic residues and quadratic nonresidues split equally, that is,*

$$|QR_p| = |QNR_p| = \frac{p-1}{2}.$$

Proof If d is a quadratic residue modulo p, then d must be among

$$\left(-\frac{p-1}{2}\right)^2, \left(-\frac{p-1}{2}+1\right)^2, \cdots, (-1)^2, 1^2, \cdots, \left(\frac{p-1}{2}-1\right)^2,$$
$$\left(\frac{p-1}{2}\right)^2 \pmod{p},$$

that is,

$$d \equiv 1^2, \cdots, \left(\frac{p-1}{2}-1\right)^2, \left(\frac{p-1}{2}\right)^2 \pmod{p}.$$

Since when $1 \le i < j \le \frac{p-1}{2}$, $i^2 \not\equiv j^2 \pmod{p}$, so the number of quadratic residues modulo p is $\frac{p-1}{2}$. The number of quadratic nonresidues is $(p-1) - \frac{p-1}{2} = \frac{p-1}{2}$.

From the proof of Theorem 3.19, we know that if the equation $x^2 \equiv d \pmod{p}$ has a solution (i.e., d is a quadratic residue modulo p), then the number of solutions is 2. Next, we present a criterion to determine whether d is a quadratic residue modulo p.

Theorem 3.20 *(Euler's criterion) Let prime $p > 2$, $(p, d) = 1$, then d is a quadratic residue modulo p if and only if*

$$d^{(p-1)/2} \equiv 1 \pmod{p};$$

d is a quadratic nonresidue modulo p if and only if

$$d^{(p-1)/2} \equiv -1 \pmod{p}.$$

Proof For any $d \in \mathbb{Z}_p^*$, by Euler's theorem,

$$d^{p-1} \equiv 1 \pmod{p},$$

so

$$(d^{(p-1)/2} - 1)(d^{(p-1)/2} + 1) \equiv 0 \pmod{p}.$$

Therefore, we have either $d^{(p-1)/2} \equiv 1 \pmod{p}$ or $d^{(p-1)/2} \equiv -1 \pmod{p}$.

Now we prove that d is a quadratic residue modulo p if and only if $d^{(p-1)/2} \equiv 1 \pmod{p}$.

Necessity: If d is a quadratic residue modulo p, then there exists x_0 such that $x_0^2 \equiv d \pmod{p}$, and by Euler's theorem we see that

$$d^{(p-1)/2} \equiv x_0^{p-1} \equiv 1 \pmod{p}.$$

Sufficiency: Consider the linear congruence equation $ax \equiv d \pmod{p}$. If a is taken to be some j in the system of reduced residues modulo p, the equation has one and only one solution $x_j \pmod{p}$. If d is not a quadratic residue modulo p, then $j \neq x_j$ and the system of reduced residues modulo p can be partitioned according to the pair j, x_j. By Wilson's theorem,

$$-1 \equiv (p-1)! \equiv (-1)^{(p-1)/2} \left(\left(\frac{p-1}{2} \right)! \right)^2 \equiv d^{(p-1)/2} \pmod{p}.$$

This contradicts our assumption, and hence, d is a quadratic residue modulo p.

The argument that d is a quadratic nonresidue modulo p is

$$d^{(p-1)/2} \equiv -1 \pmod{p}$$

can be deduced directly from the sufficient and necessary condition for d being a quadratic residue modulo p.

The theorem is proved.

The following corollaries are easy consequences of Euler's criterion.

Corollary 3.21 *If $p \equiv 1 \pmod{4}$, then -1 is a quadratic residue modulo p; if $p \equiv 3 \pmod{4}$, then -1 is a quadratic nonresidue modulo p.*

Corollary 3.22 *Let prime $p > 2$, $(p, d_1) = 1$, $(p, d_2) = 1$. Then $d_1 d_2$ is a quadratic residue modulo p if and only if both d_1, d_2 are quadratic residues modulo p or quadratic nonresidues modulo p; $d_1 d_2$ is a quadratic nonresidue modulo p if and only if one of the d_1, d_2 is a quadratic residue modulo p and the other is a quadratic nonresidue modulo p.*

Example 3.23 *Find quadratic residues and nonresidues modulo 23.*

Solution: Notice that

j	1	2	3	4	5	6	7	8	9	10	11
$d = j^2 \pmod p$	1	4	9	-7	2	-10	3	-5	-11	8	6

So $-11, -10, -7, -5, 1, 2, 3, 4, 6, 8, 9$ are quadratic residues modulo 23, and $-9, -8, -6, -4, -3, -2, -1, 5, 7, 10, 11$ are quadratic nonresidues modulo 23.

Example 3.24 *Determine the number of solutions of the following congruence equations:*

1. $x^2 \equiv 3 \pmod{91}$,

2. $x^2 \equiv 4 \pmod{55}$.

Solution: (1) The congruence equation $x^2 \equiv 3 \pmod{91}$ has the same solution with the system of congruence equations

$$\begin{cases} x^2 \equiv 3 \pmod{7}, \\ x^2 \equiv 3 \pmod{13}. \end{cases}$$

As 3 is not a quadratic residue of 7, so the system of equations has no solution. Therefore, $x^2 \equiv 3 \pmod{91}$ has no solution.

(2) The congruence equation $x^2 \equiv 4 \pmod{55}$ has the same solution with the system of congruence equations

$$\begin{cases} x^2 \equiv 4 \pmod{5}, \\ x^2 \equiv 4 \pmod{11}. \end{cases}$$

As 4 is a quadratic residue of 11 and 5, so the original equation has four solutions.

Example 3.25 *Let d be a quadratic residue modulo p. Prove that if $p \equiv 3 \pmod 4$, then $\pm d^{\frac{p+1}{4}}$ are solutions to the congruence equation $x^2 \equiv d \pmod p$.*

Proof We know that $d^{\frac{p-1}{2}} \equiv 1 \pmod p$ as d is a quadratic residue modulo p. So

$$(\pm d^{\frac{p+1}{4}})^2 = d^{\frac{p-1}{2}} \times d \equiv d \pmod p.$$

3.6 THE LEGENDRE SYMBOL AND THE JACOBI SYMBOL

In this section, we will introduce two important arithmetic functions, the Legendre symbol and the Jacobi symbol. These functions have very important applications in cryptographic algorithms.

Definition 3.26 Let prime $p > 2$, and set

$$\left(\frac{d}{p}\right) = \begin{cases} 0, & \text{if } p \mid d, \\ 1, & \text{if } d \text{ is a quadratic residue modulo } p, \\ -1, & \text{if } d \text{ is a quadratic nonresidue modulo } p. \end{cases}$$

$\left(\dfrac{d}{p}\right)$ is called the Legendre symbol modulo p. □

We have the following properties of Legendre symbol modulo p.

Theorem 3.27 *Properties of Legendre symbol:*

1. $\left(\dfrac{d}{p}\right) = d^{(p-1)/2} \pmod p$.

2. $\left(\dfrac{d}{p}\right) = \left(\dfrac{d+p}{p}\right)$.

3. $\left(\dfrac{dc}{p}\right) = \left(\dfrac{d}{p}\right)\left(\dfrac{c}{p}\right)$.

4. $\left(\dfrac{-1}{p}\right) = \begin{cases} 1, & p \equiv 1 \pmod 4, \\ -1, & p \equiv 3 \pmod 4. \end{cases}$

Proof of the theorem can be a direct consequence of the properties of congruence and the definition of the Legendre symbol; the details are left to the reader.

For a general d, the value $\left(\dfrac{d}{p}\right)$ can be computed by these proper-

ties. For special value of d, for example, $d = 2$, $\left(\dfrac{d}{p}\right)$ can be efficiently computed based on Theorem 3.29. Before proving Theorem 3.29, let us prove the following lemma.

Lemma 3.28 *Let prime $p > 2$, $(p, d) = 1$, and*

$$1 \leq j < p/2, \ t_j \equiv jd \pmod{p}, \ 0 < t_j < p.$$

Let n be the number of t_j's that are greater than $p/2$, then
$$\left(\frac{d}{p}\right) = (-1)^n.$$

Proof For any $1 \leq j < i < p/2$,
$$t_i \pm t_j \equiv (i \pm j)d \not\equiv 0 \pmod{p}.$$

That is,
$$t_i \not\equiv \pm t_j \pmod{p}.$$

Rename the t_j's that are greater than $p/2$ as $r_1, \cdots, r_n$, and the t_j's that are smaller than $p/2$ as $s_1, \cdots, s_k$. Obviously,
$$1 \leq p - r_i < p/2.$$

Since $s_j \neq p - r_i \pmod{p}$, $1 \leq j \leq k$, $1 \leq i \leq n$, so the $(p-1)/2$ numbers $s_1, \cdots, s_k, p - r_1, \cdots, p - r_n$ are exactly a permutation of $1, 2, \cdots, (p-1)/2$. By the assumption, we get

$$\begin{aligned}
&1 \times 2 \times \cdots ((p-1)/2) \times d^{(p-1)/2} \\
&\equiv t_1 t_2 \cdots t_{(p-1)/2} \\
&\equiv s_1 \cdots s_k \times r_1 \cdots r_n \equiv (-1)^n s_1 \cdots s_k \times (p - r_1) \cdots (p - r_n) \\
&\equiv (-1)^n \times 1 \times 2 \times \cdots \times ((p-1)/2) \pmod{p}.
\end{aligned}$$

Therefore,
$$\left(\frac{d}{p}\right) \equiv d^{(p-1)/2} \equiv (-1)^n \pmod{p}$$

and the lemma is proved.

From this lemma, we can get the following theorem.

Theorem 3.29 $\left(\dfrac{2}{p}\right) = (-1)^{(p^2-1)/8}$.

Proof Since $1 \le t_j = 2j < p/2$, we have $1 \le j < p/4$. From the proof of Lemma 3.28,

$$n = \frac{p-1}{2} - \left[\frac{p}{4}\right].$$

Thus,

$$n = \begin{cases} l, & p = 4l+1, \\ l+1, & p = 4l+3. \end{cases}$$

So

$$\left(\frac{2}{p}\right) = (-1)^n = \begin{cases} 1, & p \equiv \pm 1 \pmod 8, \\ -1, & p \equiv \pm 3 \pmod 8. \end{cases}$$

The theorem is proved.

Lemma 3.30 *Let prime $p > 2$. When $(d, 2p) = 1$,*

$$\left(\frac{d}{p}\right) = (-1)^T,$$

where

$$T = \sum_{j=1}^{(p-1)/2} \left[\frac{jd}{p}\right].$$

Proof Using the integer part symbol $[x]$ to $t_j \equiv jd \pmod p, 0 < t_j < p$, we have

$$jd \equiv p\left[\frac{jd}{p}\right] + t_j, \quad 1 \le j < p/2.$$

Summing both sides up with respect to j, we get

$$d \sum_{j=1}^{(p-1)/2} j = p \sum_{j=1}^{(p-1)/2} \left[\frac{jd}{p}\right] + \sum_{j=1}^{(p-1)/2} t_j = pT + \sum_{j=1}^{(p-1)/2} t_j.$$

From the proof of Lemma 3.28, we see that

$$\sum_{j=1}^{(p-1)/2} t_j = s_1 + \cdots + s_k + r_1 + \cdots + r_n$$

$$= s_1 + \cdots + s_k + (p - r_1) + \cdots + (p - r_n) - np$$

$$+ 2(r_1 + \cdots + r_n)$$

$$= \sum_{j=1}^{(p-1)/2} j - np + 2(r_1 + \cdots + r_n).$$

These two relations give

$$\frac{p^2 - 1}{8}(d - 1) = p(T - n) + 2(r_1 + \cdots + r_n).$$

The proof is complete.

Theorem 3.31 *(Gauss' Law of Quadratic Reciprocity) Let p, q be odd primes, $p \neq q$, then*

$$\left(\frac{q}{p}\right)\left(\frac{p}{q}\right) = (-1)^{(p-1)/2 \cdot (q-1)/2}.$$

Proof By Lemma 3.30, we get

$$\left(\frac{q}{p}\right)\left(\frac{p}{q}\right) = (-1)^{S+T},$$

where

$$S = \sum_{i=1}^{(p-1)/2}\left[\frac{iq}{p}\right], \quad T = \sum_{j=1}^{(q-1)/2}\left[\frac{jp}{q}\right].$$

In fact, S is the number of integral points in the interior (excluding the boundary) of the region enclosed by three straight lines $y = \frac{q}{p}x$, $x = \frac{p}{2}$, $y = 0$, while T is the number of integral points in the interior of the region enclosed by three straight lines $y = \frac{p}{q}x$, $x = \frac{q}{2}$, $y = 0$. A slight modification shows that T is also the number of integral points in the interior of the region enclosed by three straight lines $y = \frac{q}{p}x$, $y = \frac{q}{2}$, $x = 0$. Therefore, $S + T$ becomes the number of integral points in the interior of the rectangle $x = \frac{p}{2}$, $y = 0$, $x = 0$, $y = \frac{q}{2}$ (the line $y = \frac{q}{p}x$ does not pass any integral points within this region), namely,

$$S + T = \frac{p-1}{2}\frac{q-1}{2}.$$

The theorem is proved.

Example 3.32 *Evaluate* $\left(\dfrac{157}{751}\right)$.

Solution: From Gauss' Law of Quadratic Reciprocity,

$$\left(\frac{157}{751}\right) = \left(\frac{751}{157}\right).$$

From the properties of Legendre symbol,

$$\left(\frac{751}{157}\right) = \left(\frac{-34}{157}\right) = \left(\frac{-1}{157}\right)\left(\frac{2}{157}\right)\left(\frac{17}{157}\right).$$

Calculation shows

$$\left(\frac{-1}{157}\right) = 1, \left(\frac{2}{157}\right) = -1, \left(\frac{17}{157}\right) = \left(\frac{157}{17}\right) = \left(\frac{4}{17}\right) = 1.$$

So

$$\left(\frac{157}{751}\right) = -1.$$

Example 3.33 *Find all odd primes for which 7 is a quadratic residue.*

Solution: From Gauss' Law of Quadratic Reciprocity,

$$\left(\frac{7}{p}\right) = (-1)^{(p-1)/2}\left(\frac{p}{7}\right).$$

A direct calculation shows

$$\left(\frac{p}{7}\right) = \begin{cases} 1, & p \equiv 1,2,-3 \pmod 7, \\ -1, & p \equiv -1,-2,3 \pmod 7. \end{cases} \tag{3.8}$$

$$(-1)^{(p-1)/2} = \begin{cases} 1, & p \equiv 1 \pmod 4, \\ -1, & p \equiv 3 \pmod 4. \end{cases} \tag{3.9}$$

Consider the congruence equation

$$\begin{cases} x \equiv a_1 \pmod 4, \\ x \equiv a_2 \pmod 7, \end{cases}$$

where when $a_1 = 1$, a_2 is taken to be $1, 2, -3$; when $a_1 = -1$, a_2 is taken to be $-1, -2, 3$. By the Chinese remainder theorem, when

$$p \equiv \pm 1, \pm 3, \pm 9 \pmod{28},$$

$\left(\frac{7}{p}\right) = 1$. That is, 7 is a quadratic residue modulo p.

Based on the Legendre symbol, we define an arithmetic function that is of more general form—the Jacobi symbol.

Definition 3.34 Let odd number $p > 1$, $P = p_1 p_2 \cdots p_n$, where $p_i(1 \le i \le n)$ are prime numbers. We call

$$\left(\frac{d}{P}\right) = \left(\frac{d}{p_1}\right)\left(\frac{d}{p_2}\right)\cdots\left(\frac{d}{p_n}\right)$$

the **Jacobi symbol.** Here $\left(\dfrac{d}{p_i}\right), (1 \le i \le n)$ are Legendre symbols. □

Similar to the Legendre symbol, the Jacobi symbol has the following properties.

Theorem 3.35 *The Jacobi symbol has the following properties:*

1. $\left(\dfrac{1}{P}\right) = 1.$

2. $(d, P) \ne 1, \left(\dfrac{d}{P}\right) = 0.$

3. $(d, P) = 1, \left(\dfrac{d}{P}\right) = \pm 1.$

4. $\left(\dfrac{d}{P}\right) = \left(\dfrac{d+P}{P}\right).$

5. $\left(\dfrac{dc}{P}\right) = \left(\dfrac{d}{P}\right)\left(\dfrac{c}{P}\right).$

These properties are direct consequences of the definition of the Jacobi symbol.

Theorem 3.36

1. $\left(\dfrac{-1}{P}\right) = (-1)^{(P-1)/2}.$

2. $\left(\dfrac{2}{P}\right) = (-1)^{(P^2-1)/8}.$

Proof Let $a_i \equiv 1 (\bmod\ m)(1 \le i \le s)$, $a = a_1 \cdots a_s$, then we can prove the following equality:

$$\frac{a-1}{m} = \frac{a_1 - 1}{m} + \cdots + \frac{a_s - 1}{m} (\bmod\ m).$$

In fact, we only need to prove the case of $s = 2$; the rest would be similar. First, from $a = a_1 a_2$, we see that

$$a - 1 = a_1 a_2 - 1 = (a_1 - 1) + (a_2 - 1) + (a_1 - 1)(a_2 - 1).$$

The fact $a_i \equiv 1 \pmod{m}$ implies that $a \equiv 1 \pmod{m}$; therefore,

$$\frac{a-1}{m} = \frac{a_1 - 1}{m} + \frac{a_2 - 1}{m} + \frac{(a_1 - 1)(a_2 - 1)}{m}$$
$$= \frac{a_1 - 1}{m} + \frac{a_2 - 1}{m} \pmod{m}.$$

Take $m = 2$, $a_i = p_i (1 \le i \le n)$, $a = P$, we see that

$$\left(\frac{-1}{P}\right) = \left(\frac{-1}{p_1}\right) \cdots \left(\frac{-1}{p_n}\right) = (-1)^{(p_1-1)/2 + \cdots + (p_n-1)/2} = (-1)^{(P-1)/2}.$$

Similarly, take $m = 8$, $a_i = p_i (1 \le i \le n)$, $a = P$, we get

$$\left(\frac{2}{P}\right) = (-1)^{(P^2-1)/8}.$$

The theorem is proved.

The Jacobi symbol also obeys the Law of Quadratic Reciprocity.

Theorem 3.37 Let $P > 1$, $Q > 1$ be odd numbers with $(Q, P) = 1$, then

$$\left(\frac{Q}{P}\right)\left(\frac{P}{Q}\right) = (-1)^{(P-1)/2 \cdot (Q-1)/2}.$$

Proof Let $Q = q_1 \cdots q_s$, $P = p_1 \cdots p_t$, then

$$\left(\frac{Q}{P}\right) = \prod_{i=1}^{t}\prod_{j=1}^{s}\left(\frac{q_j}{p_i}\right) = \prod_{i=1}^{t}\prod_{j=1}^{s}\left(\frac{p_i}{q_j}\right)(-1)^{(p_i-1)/2 \cdot (q_j-1)/2}$$

$$= \left\{\prod_{i=1}^{t}\prod_{j=1}^{s}\left(\frac{p_i}{q_j}\right)\right\}\left\{\prod_{i=1}^{t}\prod_{j=1}^{s}(-1)^{(p_i-1)/2 \cdot (q_j-1)/2}\right\}.$$

From the proof of Theorem 3.36, we see that

$$\frac{Q-1}{2} \equiv \sum_{j=1}^{s}(q_j - 1)/2 \pmod 2, \quad \frac{P-1}{2} \equiv \sum_{i=1}^{t}(p_i - 1)/2 \pmod 2.$$

Thus,

$$\left(\frac{Q}{P}\right)\left(\frac{P}{Q}\right) = (-1)^{(P-1)/2 \cdot (Q-1)/2}.$$

Theorem 3.37 and the properties of the Jacobi symbol can be used to compute Jacobi symbols of any form. This is in fact another important application of the Euclidean algorithm. The Legendre symbol can be computed as a Jacobi symbol.

Remark A difference between the Legendre symbol and the Jacobi symbol is that the Jacobi symbol $\left(\frac{d}{P}\right) = 1$ does not necessarily mean that the quadratic congruence equation $x^2 \equiv d \pmod P$ must have a solution.

Example 3.38 *Find the Jacobi symbol:* $\left(\dfrac{567}{783}\right)$.

Solution: Since $(567, 783) = 27 > 1$, so $\left(\dfrac{567}{783}\right) = 0$.

EXERCISES

3.1 Solve the following congruence equations:

(1) $4x^2 + 27x - 12 \equiv 0 \pmod{15}$.

(2) $x^2 + 3x - 5 \equiv 0 \pmod{13}$.

3.2 Use identical transformation to solve the following congruence equations:

(1) $x^7 + 6x^6 - 13x^4 - x^3 - 2x^2 + 40x - 9 \equiv 0 \pmod{5}$.

(2) $x^9 - 4x^8 - 5x^7 + x^2 + 5x + 2 \equiv 0 \pmod{7}$.

3.3 Let p be a prime, if $g(x) \equiv 0 \pmod{p}$ has no solution, the $f(x) \equiv 0 \pmod{p}$ and $f(x)g(x) \equiv 0 \pmod{p}$ have the same number of solutions.

3.4 Solve the following linear congruence equations:

(1) $8x \equiv 6 \pmod{10}$.

(2) $3x \equiv 10 \pmod{17}$.

3.5 Solve the following systems of congruence equations:

(1) $\begin{cases} x \equiv 3 & \pmod 7, \\ x \equiv 5 & \pmod{11}. \end{cases}$

(2) $\begin{cases} x \equiv 2 & \pmod 8, \\ x \equiv 6 & \pmod{11}, \\ x \equiv -1 & \pmod{21}. \end{cases}$

3.6 Let $(a, b) = 1, c \neq 0$. Prove that there must be a positive integer n such that $(a + bn, c) = 1$.

3.7 Solve the congruence equation modulo power of primes:

(1) $x^2 + 2x + 1 \equiv 0 \pmod{3^2}$.

(2) $x^2 + 5x + 13 \equiv 0 \pmod{3^3}$.

3.8 Let $T(m; f)$ be the number of solutions of the congruence equation $f(x) \equiv 0 \pmod m$ with $(x, m) = 1$. Prove that if $(m_1, m_2) = 1$, then $T(m_1 m_2; f) = T(m_1; f)T(m_2; f)$.

3.9 Find the number of solutions of the following congruence equations:

(1) $x^2 \equiv 43 \pmod{109}$.

(2) $x^2 \equiv 7 \pmod{83}$.

3.10 Let $p \equiv 1 \pmod 4$ be an odd prime. Prove that

    (1) The numbers of quadratic residues and quadratic non-residues modulo p in $1, 2, \cdots, (p-1)/2$ are both $(p-1)/4$.

    (2) In $1, 2, \cdots, (p-1)$, there are $(p-1)/8$ even numbers that are quadratic residues modulo p and $(p-1)/8$ odd numbers that are quadratic nonresidues modulo p.

3.11 Use the properties of the Jacobi symbol to compute $\left(\dfrac{205}{8633} \right)$.

3.12 Find all odd primes p for which 19 is a quadratic residue modulo p.

3.13 If prime $p > 2$, prove that $x^4 \equiv -4 \pmod p$ has a solution if and only if $p \equiv 1 \pmod 4$.

3.14 Let $p \geq 3, p \nmid a$. Prove that

$$\sum_{x=1}^{p} \left(\frac{x^2 + ax}{p} \right) = -1.$$

3.15 Let $n = p_1 p_2 p_3$, where $p_i, i = 1, 2, 3$ are distinct primes. Set

$$J^{+1} = \left\{ x \,\middle|\, x \in Z_n^*, \left(\frac{x}{n} \right) = 1 \right\}, \quad J^{-1} = \left\{ x \,\middle|\, x \in Z_n^*, \left(\frac{x}{n} \right) = -1 \right\},$$

where $\left(\dfrac{x}{n} \right)$ denotes the Jacobi symbol, and Z_n^* denotes the reduced system of residues modulo n. Complete the following:

    (1) Prove that $|J^{+1}| = |J^{-1}|$, where $|J^{+1}|$, $|J^{-1}|$ denotes the number of elements of J^{+1}, J^{-1}, respectively.

    (2) Compute the probability that x is a quadratic residue modulo n, given that $x \in J^{+1}$.

    (3) Compute the probability that xy is a quadratic residue modulo n, given that $x, y \in J^{+1}$.

CHAPTER 4

Exponents and Primitive Roots

I N A REDUCED system of residues, if the exponent of an element is exactly $\varphi(m)$, then this element is called a primitive root modulo m. In a reduced system of residues with a primitive root, every element can be expressed as a power of the primitive root. Conversely, all elements represented by powers of a primitive root form a reduced system of residues. This gives a natural method for constructing a reduced system of residues. However, primitive roots exist only for $m = 1, 2, 4, p^{\alpha}, 2p^{\alpha}$. How do we construct a reduced system of residues for modulus m for which the primitive root does not exist? All of these will be discussed in this chapter. We will also introduce two major concepts, the exponents and indexes, and their properties. Indexes are called discrete logarithms in cryptography, and the discrete logarithm problem is an important theoretical base for the design of several public key cryptographic algorithms.

4.1 EXPONENTS AND THEIR PROPERTIES

In this section, we present the concepts of exponents and primitive roots, as well as their properties.

Definition 4.1 Let $m \geq 1$, $(a, m) = 1$. The smallest positive integer d such that $a^d \equiv 1 \pmod{m}$ is called an exponent of a modulo m (it is also called an order, or a period), denoted by $\delta_m(a)$. If $\delta_m(a) = \varphi(m)$, then a is called a primitive root modulo m. □

Property 1 Let $m \geq 1$ and $(a, m) = 1$. For any integer d, if $a^d \equiv 1 \pmod{m}$, then $\delta_m(a)|d$.

Proof Let $d_0 = \delta_m(a)$. By the division algorithm, there exist q, r such that $d = qd_0 + r$, $0 \leq r < d_0$. Therefore,

$$a^d - 1 = a^{qd_0+r} - 1 = (a^{d_0})^q a^r - 1 \equiv a^r - 1 \equiv 0 \pmod{m}.$$

Since $0 \leq r < d_0$, by the definition of exponent, we have $r = 0$. The result is proved.

Property 2 If $b \equiv a \pmod{m}$, $(a, m) = 1$, then $\delta_m(a) = \delta_m(b)$.

Property 3 $\delta_m(a)|\varphi(m)$ and $\delta_{2^l}(a)|2^{l-2}$ for $l \geq 3$.

Property 3 can be used to verify the correctness of the following examples.

Example 4.2 *List the exponents of all elements in the reduced system of residues modulo $m = 17$.*

a	1	2	3	4	5	6	7	8	9	10	11	12	13	14	15	16
$\delta(a)$	1	8	16	4	16	16	16	8	8	16	16	16	4	16	8	2

From $\varphi(m) = 16$ and the above table, we know that the primitive roots modulo 17 are $3, 5, 6, 7, 10, 11, 12, 14 \pmod{17}$.

Example 4.3 *List the exponents of all elements in the reduced system of residues modulo $m = 2^5$.*

a	1	3	5	7	9	11	13	15	17	19	21	23	25	27	29	31
$\delta(a)$	1	8	8	4	4	8	8	2	2	8	8	4	4	8	8	2

From $\varphi(m) = 2^4 = 16$ and the above table, we know that there is no primitive root modulo $m = 2^5$.

Property 4 If $(a, m) = 1$, $a^i \equiv a^j \pmod{m}$, then $i \equiv j \pmod{\delta_m(a)}$.

Property 5 If $aa^{-1} \equiv 1 \pmod{m}$, then $\delta_m(a) = \delta_m(a^{-1})$.

The proofs of Properties 2, 3, 4, and 5 are straightforward and are left as exercises.

Property 6 Let k be a nonnegative integer, then

$$\delta_m(a^k) = \frac{\delta_m(a)}{(\delta_m(a), k)}.$$

Furthermore, in a reduced system of residues modulo m, there are at least $\varphi(\delta_m(a))$ elements whose exponents modulo m are $\delta_m(a)$.

Proof Let $\delta = \delta_m(a)$, $\delta' = \delta/(\delta, k)$, $\delta'' = \delta_m(a^k)$. We first prove that $\delta' | \delta''$. Since

$$(a^k)^{\delta''} \equiv 1 \pmod{m},$$

by property 1, we get $\delta | k\delta''$, and hence, $\delta' = \dfrac{\delta}{(\delta, k)} \mid \dfrac{k\delta''}{(\delta, k)}$. As $\left(\dfrac{\delta}{(\delta, k)}, \dfrac{k}{(\delta, k)} \right) = 1$, so $\delta' | \delta''$ holds.

Next, we prove that $\delta'' | \delta'$. From $a^{k\delta'} \equiv (a^k)^{\delta'} \equiv 1 \pmod{m}$, we know that $\delta'' | \delta'$ holds. Therefore, $\delta' = \delta''$ and the proof is complete.

The following two important corollaries follow from property 6.

Corollary 4.4 *When* $(k, \delta_m(a)) = 1$, $\delta_m(a) = \delta_m(a^k)$.

Corollary 4.4 is an important result because it not only determines the primitive roots and as well as the number of primitive roots, but it can also be used to give the generators and the number of generators in a finite cyclic group.

Corollary 4.5 *If g is a primitive root modulo m, then the number of primitive roots modulo m is $\varphi(\varphi(m))$. Furthermore,*

$$\{g^i | (i, \varphi(m)) = 1, 1 \le i < \varphi(m)\}$$

is the set of all primitive roots modulo m.

Property 7 $\delta_m(ab) = \delta_m(a)\delta_m(b)$ if and only if $(\delta_m(a), \delta_m(b)) = 1$.

Proof Let $\delta = \delta_m(ab)$, $\delta' = \delta_m(a)$, $\delta'' = \delta_m(b)$, $\eta = [\delta_m(a), \delta_m(b)]$. First, we note that

$$1 = (ab)^\delta \equiv (ab)^{\delta\delta''} \equiv a^{\delta\delta''} \pmod{m}.$$

So $\delta' | \delta\delta''$. Since $(\delta', \delta'') = 1$, we get $\delta' | \delta$. Similarly,

$$1 \equiv (ab)^\delta \equiv (ab)^{\delta\delta'} \equiv b^{\delta\delta'} \pmod{m},$$

so $\delta'' | \delta\delta'$, and hence, $\delta'' | \delta$. Using the fact $(\delta', \delta'') = 1$ again, we get $\delta'\delta'' | \delta$. It is obvious that $(ab)^{\delta'\delta''} \equiv 1 \pmod{m}$, so $\delta | \delta'\delta''$ holds. Therefore, $\delta = \delta'\delta''$, and the sufficiency is proved.

For necessity, we note that $(ab)^\eta \equiv 1 \pmod{m}$, so $\delta | \eta$. From $\delta = \delta'\delta''$, we get $\delta'\delta'' | \eta$. Note that it is obvious that $\eta | \delta'\delta''$, so $\delta'\delta'' = \eta$ holds true. Therefore, $(\delta', \delta'') = 1$.

Property 8

1. If $n|m$, then $\delta_n(a)|\delta_m(a)$.

2. If $(m_1, m_2) = 1$, then $\delta_{m_1 m_2}(a) = [\delta_{m_1}(a), \delta_{m_2}(a)]$.

Proof

1. From $a^{\delta_m(a)} \equiv 1 \pmod{m}$, we get $a^{\delta_m(a)} \equiv 1 \pmod{n}$, so $\delta_n(a)|\delta_m(a)$.

2. Let $\delta' = [\delta_{m_1}(a), \delta_{m_2}(a)]$. Since $\delta_{m_1}(a)|\delta_{m_1 m_2}(a), \delta_{m_2(a)}|\delta_{m_1 m_2}(a)$, so $\delta'|\delta_{m_1 m_2}(a)$. On the other hand, $a^{\delta'} \equiv 1 \pmod{m_j}$, $(j = 1, 2)$, and thus, $(m_1, m_2) = 1$ yields $a^{\delta'} \equiv 1 \pmod{m_1 m_2}$, so we get $\delta_{m_1 m_2}(a)|\delta'$.

From property 8, a more general property (i.e., property 9) can be obtained.

Property 9 If $m = 2^{\alpha_0} p_1^{\alpha_1} p_2^{\alpha_2} \cdots p_r^{\alpha_r}$, where $p_i(1 \le i \le r)$ are distinct odd prime numbers, then $\delta_m(a)|\lambda(m)$, where

$$\lambda(m) = [2^{c_0}, \varphi(p_1^{\alpha_1}), \cdots, \varphi(p_r^{\alpha_r})], \quad c_0 = \begin{cases} 0, & \alpha_0 = 0, 1; \\ 1, & \alpha_0 = 2; \\ \alpha_0 - 2, & \alpha_0 \ge 3. \end{cases}$$

$\lambda(m)$ is called the Carmichael function.

Property 10 Let $(m_1, m_2) = 1$. Then for any a_1, a_2, there must be an a such that

$$\delta_{m_1 m_2}(a) = [\delta_{m_1}(a_1), \delta_{m_2}(a_2)].$$

Proof Consider the system of congruence equations:

$$x \equiv a_1 \pmod{m_1}, \ x \equiv a_2 \pmod{m_2}.$$

By the Chinese remainder theorem, this system has a unique solution

$$x \equiv a \pmod{m_1 m_2}.$$

It is obvious that $\delta_{m_1}(a) = \delta_{m_1}(a_1)$, $\delta_{m_2}(a) = \delta_{m_2}(a_2)$. Therefore, the result follows from property 8.

Property 11 For any a, b, there must be a c such that

$$\delta_m(c) = [\delta_m(a), \delta_m(b)].$$

Proof Let $\delta' = \delta_m(a)$, $\delta'' = \delta_m(b)$, $\eta = [\delta', \delta'']$. We can factorize δ', δ'' as follows:

$$\delta' = \tau'\eta', \delta'' = \tau''\eta'',$$

where

$$(\eta', \eta'') = 1, \eta'\eta'' = \eta.$$

From property 6, we get

$$\delta_m(a^{\tau'}) = \eta', \delta_m(b^{\tau''}) = \eta''.$$

From property 7, we get

$$\delta_m(a^{\tau'}b^{\tau''}) = \delta_m(a^{\tau'})\delta_m(b^{\tau''}) = \eta'\eta'' = \eta.$$

Therefore, the result is proved by taking $c = a^{\tau'}b^{\tau''}$.

Example 4.6 *Let $m > 1$, $(ab, m) = 1$, and λ be the smallest positive integer d such that $a^d \equiv b^d \pmod{m}$. Prove that*

1. *If $a^k \equiv b^k \pmod{m}$, then $\lambda \mid k$.*

2. *$\lambda \mid \varphi(m)$.*

Proof

1. Let $k = \lambda q + r$, $0 \le r < \lambda$:

$$a^k = a^{\lambda q + r} = a^{\lambda q}a^r \equiv b^k \equiv b^{\lambda q + r} \equiv b^{\lambda q}b^r \pmod{m}.$$

   As $a^\lambda \equiv b^\lambda \pmod{m}$, and $(ab, m) = 1$, by properties of congruence we get
   $$a^r \equiv b^r \pmod{m}.$$

   Since λ is the smallest d with $a^d \equiv b^d \pmod{m}$, we must have $r = 0$, that is, $\lambda \mid k$.

2. Since $(ab, m) = 1$, $a^{\varphi(m)} \equiv 1 \equiv b^{\varphi(m)} \pmod{m}$, we get $\lambda \mid \varphi(m)$ by the discussion above.

4.2 PRIMITIVE ROOTS AND THEIR PROPERTIES

The next theorem states a sufficient and necessary condition for modulus m to possess a primitive root.

Theorem 4.7 *A modulus m has a primitive root if and only if $m = 1, 2, 4, p^\alpha, 2p^\alpha$, where p is odd prime and $\alpha \geq 1$.*

Proof of the necessity of the theorem. If m is not a number listed in the theorem, we must have

$$m = 2^\alpha, \ (\alpha \geq 3), \ \text{or} \ m = 2^{\alpha_0} p_1^{\alpha_1} p_2^{\alpha_2} \cdots p_r^{\alpha_r}, \ (\alpha_0 \geq 2, r \geq 1).$$

or

$$m = 2^{\alpha_0} p_1^{\alpha_1} p_2^{\alpha_2} \cdots p_r^{\alpha_r}, \ (\alpha_0 \geq 0, r \geq 2),$$

where p_i are distinct odd primes, $\alpha_i \geq 1 (1 \leq i \leq r)$. Let

$$\lambda(m) = [2^{c_0}, \varphi(p_1^{\alpha_1}), \cdots, \varphi(p_r^{\alpha_r})], \ c_0 = \begin{cases} 0, & \alpha_0 = 0, 1; \\ 1, & \alpha_0 = 2; \\ \alpha_0 - 2, & \alpha_0 \geq 3; \end{cases}$$

then it is easy to check that when m is one of the three cases above, we always have $\lambda(m) < \varphi(m)$. From the properties discussed in Section 4.1, $\delta_m(a) \leq \lambda(m)$; hence, $\delta_m(a) < \varphi(m)$, and the modulus does not have a primitive root.

We need two lemmas before proving the sufficiency of the theorem.

Lemma 4.8 *Let p be a prime, then there exists a primitive roots modulo p.*

Proof From property 11 of Section 4.1, there must be an integer g such that

$$\delta_p(g) = [\delta_p(1), \delta_p(2), \cdots, \delta_p(p-1)] = \delta.$$

Obviously, $\delta | p - 1$, so $\delta \leq p - 1$. As $\delta_p(i) | \delta$ for $i = 1, 2, \cdots, p - 1$, we see that $1, 2, \cdots, p - 1 \pmod{p}$ are solutions of the congruence equation

$$x^\delta \equiv 1 \pmod{p}.$$

Note that the number of solutions of this congruence equation $n \leq \min\{\delta, p\}$, so $p - 1 \leq \delta$. This shows that $\delta = p - 1$, and g is a primitive root modulo p.

Lemma 4.9 *Let p be an odd prime. Then for any $\alpha \geq 1$, there exist primitive roots modulo p^α, $2p^\alpha$.*

Proof This lemma is proved through the next five steps.

1. If g is a primitive root modulo $p^{\alpha+1}$, then g must be a primitive root modulo p^α.

To this end, it is sufficient to show $\delta_{p^\alpha}(g) = \varphi(p^\alpha)$. Let $\delta = \delta_{p^\alpha}(g)$, then $\delta|\varphi(p^\alpha)$ and $g^{p\delta} \equiv 1 \pmod{p^{\alpha+1}}$. We see that from the fact that g is a primitive root modulo $p^{\alpha+1}$,

$$\varphi(p^{\alpha+1}) = \delta_{p^{\alpha+1}}(g)|p\delta.$$

However, since $\varphi(p^{\alpha+1}) = p^\alpha(p-1)$, so $\varphi(p^\alpha)|\delta$. Therefore, $\delta = \varphi(p^\alpha)$. That is, g must be a primitive root modulo p^α.

2. If g is a primitive root modulo p^α, then we must have $\delta_{p^{\alpha+1}}(g) = \varphi(p^\alpha)$ or $\varphi(p^{\alpha+1})$.

As $p^\alpha|p^{\alpha+1}$, property 8 from Section 4.1 implies

$$\varphi(p^\alpha) = \delta_{p^\alpha}(g)|\delta_{p^{\alpha+1}}(g).$$

However, since

$$\delta_{p^{\alpha+1}}(g)|\varphi(p^{\alpha+1}),$$

so

$$\delta_{p^{\alpha+1}}(g) = \varphi(p^\alpha) \text{ or } \varphi(p^{\alpha+1}).$$

3. Suppose p is an odd prime, if g is a primitive root modulo p and $g^{p-1} = 1 + rp$ with $(p, r) = 1$, then g is a primitive root modulo p^α ($\alpha \geq 1$).

We first prove for $\alpha \geq 1$, $g^{\varphi(p^\alpha)} = 1 + r(\alpha)p^\alpha$, $(p, r(\alpha)) = 1$ holds.

The result is true when $\alpha = 1$. Now assume the result is true for $\alpha = n$ ($n \geq 1$). Then for $\alpha = n+1$,

$$g^{\varphi(p^{n+1})} = (1 + r(n)p^n)^p$$
$$= 1 + r(n)p^{n+1} + \frac{1}{2}p(p-1)r^2(n)p^{2n} + \cdots$$
$$= 1 + r(n+1)p^{n+1}.$$

Since $(p, r(n)) = 1$, so $(p, r(n+1)) = 1$, this means that the result is true for $\alpha = n+1$. This, together with the discussion above, implies that g is a primitive root modulo p^α ($\alpha > 1$).

4. Suppose that p is an odd prime, g' is an odd primitive root modulo p (if g' is even, then replace it by $g' + p$). Then for each $t = 0, 1, \cdots, p-1$, $g = g' + tp$ is a primitive root modulo p; moreover, all but one of these numbers satisfy $g^{p-1} = 1 + rp$ with $(p, r) = 1$.

So

$$g^{p-1} = (g' + tp)^{p-1} = (g')^{p-1} + (p-1)(g')^{p-2}pt + Ap^2,$$

where A is an integer. Write $(g')^{p-1} = 1 + ap$, then

$$g^{p-1} = 1 + ((p-1)(g')^{p-2}t + a)p + Ap^2.$$

Note that $(p, (p-1)g') = 1$, so the congruence equation

$$(p-1)(g')^{p-2}t + a \equiv 0 \pmod{p}$$

has only one solution in t. This proves what we wanted.

We can always choose an odd primitive root modulo p such that $g^{p-1} = 1 + rp$, $(p, r) = 1$, as there are at least two even numbers in $t = 0, 1, \cdots, p-1$. We denote such a primitive root as $\bar{g}$.

5. From steps 3 and 4, it is immediate that $\bar{g}$ is a primitive root modulo all p^α. Since $\bar{g}$ is an odd number,

$$(\bar{g})^d \equiv 1 \pmod{p^\alpha} \Leftrightarrow (\bar{g})^d \equiv 1 \pmod{2p^\alpha}.$$

Therefore, $\delta_{2p^\alpha}(\bar{g}) = \delta_{p^\alpha}(\bar{g}) = \varphi(p^\alpha)$. Together with the fact $\varphi(2p^\alpha) = \varphi(p^\alpha)$, we get that $\bar{g}$ is a primitive root modulo all $2p^\alpha (\alpha \geq 1)$. In fact, there exists a $\bar{g}$ such that for all $\alpha \geq 1$, $\bar{g}$ is a common primitive root modulo p^α and $2p^\alpha$.

Proof of the sufficiency of the theorem. By Lemmas 4.8 and 4.9, for odd prime p, there exists primitive root modulo $m = p, p^\alpha, 2p^\alpha$. For $m = 1, 2, 4$, it is easy to verify that the corresponding primitive roots are $1, 1, -1$, respectively. Therefore, when $m = 1, 2, 4, p^\alpha, 2p^\alpha$ (p is odd prime), there exists a primitive root modulo m. The theorem is proved.

According to Theorem 4.7, the method of finding a primitive root modulo $m = p$ is quite complicated. In general, it requires us to factorize $\varphi(m)$ and verify $a^d \not\equiv 1 \pmod{m}$ for all proper divisors d of $\varphi(m)$. In fact, finding a concrete primitive root is indeed a difficult problem. There is no general method available. Theorem 4.10 provides a relatively simple way of searching primitive roots. It is the most commonly used method for finding primitive roots in cryptography.

Theorem 4.10 *Let $m = 1, 2, 4, p^\alpha, 2p^\alpha$ (with p being an odd prime). Suppose that all distinct prime divisors of $\varphi(m)$ are $q_1, q_2, \cdots, q_s$, then g is a primitive root modulo m if and only if*

$$g^{\varphi(m)/q_j} \not\equiv 1 \pmod{m}, \ j = 1, \cdots, s.$$

For a randomly selected integer a, we can determine whether a is a primitive root modulo m by Theorem 4.10. From Corollary 4.5 of Section 4.1, we know that the average probability of distribution of primitive roots is $\varphi(\varphi(m))/p$. This probability indicates that we can get a primitive root in polynomial time by randomized methods. Theorem 4.10 gives a commonly used probabilistic method for finding primitive roots in cryptography.

Example 4.11 *Find a primitive root modulo $p = 61$.*

Solution: $p = 61$, $\varphi(p) = 60 = 2^2 \times 3 \times 5$. Since

$$2^{60/5} = 2^{12} \not\equiv 1 \pmod{61}, 2^{60/3} = 2^{20} \not\equiv 1 \pmod{61},$$
$$2^{60/2} = 2^{30} \not\equiv 1 \pmod{61}.$$

So 2 is a primitive root modulo 61.

4.3 INDICES, CONSTRUCTION OF REDUCED SYSTEM OF RESIDUES

An index is a basic concept in elementary number theory. The problem of finding an index is exactly the problem of finding a discrete logarithm. We use $index_{m,g}(a)$ to denote an index; in cryptography, this is usually called a discrete logarithm.

Theorem 4.12 *If there exists a primitive root modulo m, then any primitive root g generates a reduced system of resides modulo m. More specifically, $\{g^0, g^1, \cdots, g^{\varphi(m)-1}\}$ is the reduced system of residues modulo m.*

Proof By the definition of primitive roots, $\varphi(m)$ is the smallest number d such that $g^d \equiv 1 \pmod{m}$. Therefore, it is easy to see that for $0 \le i < \varphi(m), 0 \le j < \varphi(m)$, when $i \ne j$,

$$g^i \not\equiv g^j \pmod{m}.$$

The theorem is proved.

We usually say that a primitive root g is a generator of the reduced system of residues modulo m. This is consistent with the generator of a finite cyclic group.

Definition 4.13 Let g be a primitive root modulo m. Given a with $(a, m) = 1$, there exists a unique $\gamma, 0 \le \gamma < \varphi(m)$ such that $a \equiv g^\gamma$ (mod m). We call γ the index (or logarithm) of a modulo m with base g, and it is denoted by $\gamma_{m,g}(a)$ or $index_{m,g}(a)$. If the modulus m and the primitive root g are clear in the context, it can simply be written as $\gamma_g(a)$, $\gamma(a)$ or $index_g(a)$, $index(a)$. $\square$

The next theorem proves that in the reduced system of residues modulo $m = 2^\alpha$, there must be an element g_0 that satisfies $\delta_{2^\alpha}(g_0) = 2^{\alpha-2}$. More precisely, we can choose g_0 to be 5.

Theorem 4.14 Let $m = 2^l$ $(l > 3)$, $a = 5$. Then the smallest positive integer d such that $a^d \equiv 1$ (mod m) is 2^{l-2}.

Proof Note that $\varphi(2^l) = 2^{l-1}$. From $d | \varphi(2^l)$, we know that $d = 2^k$, $0 \le k \le l - 1$.

We first prove that for any odd number a,

$$a^{2^{l-2}} \equiv 1 \pmod{2^l}.$$

In fact, let $a = 2t + 1$. If $l = 3$, then $a^2 = 4t(t+1) + 1 \equiv 1 \pmod{2^3}$, and the result is true. Assume that the result is true when $l = n$ $(n \ge 3)$. Now consider $l = n + 1$. By

$$a^{2^{n-1}} - 1 = (a^{2^{n-2}} - 1)(a^{2^{n-2}} + 1)$$

and the induction hypothesis $a^{2^{n-2}} \equiv 1 \pmod{2^n}$, we see that the result is true when $l = n + 1$.

Next, we show that when $a = 5$,

$$5^{2^{l-3}} \not\equiv 1 \pmod{2^l}$$

must hold for any $l \ge 3$.

This can be directly checked for $l = 3$. Assume that the result is true for $l = n$ $(n \ge 3)$. Then when $l = n + 1$, we have

$$5^{2^{l-3}} \equiv 1 \pmod{2^{l-1}}, \ l \ge 3.$$

Therefore,

$$5^{2^{n-3}} = 1 + s \cdot 2^{n-1}, \ 2 \nmid s.$$

Hence,
$$5^{2^{n-2}} = 1 + s(1 + s \cdot 2^{n-2})2^n, \ 2 \nmid s(1 + s \cdot 2^{n-2}).$$

This proves that the result also holds when $l = n + 1$.

In summary, we know that for $a = 5$, the smallest positive integer d such that $a^d \equiv 1 \pmod{m}$ is 2^{l-2}.

Definition 4.15 In the reduced system of residues modulo $m = 2^\alpha$, $\alpha \geq 3$, if there is a g_0 such that $\delta_{2^\alpha}(g_0) = 2^{\alpha-2}$, then

$$\pm g_0^0, \pm g_0^1, \cdots, \pm g_0^{2^{\alpha-2}-1}$$

form a reduced system of residues modulo $m = 2^\alpha$. So any odd number a can be represented uniquely as

$$a \equiv (-1)^{\gamma^{(-1)}} g_0^{\gamma^{(0)}} \pmod{2^\alpha}, \ 0 \leq \gamma^{(-1)} < 2, \ 0 \leq \gamma^{(0)} < 2^{\alpha-2}.$$

We call $\gamma^{(-1)}, \gamma^{(0)}$ the index tuple of a modulo 2^α with the base $-1, g_0$. They are denoted as $\gamma_{2^\alpha,-1,g_0}^{(-1)}(a), \gamma_{2^\alpha,-1,g_0}^{(0)}(a)$, or simply as $\gamma^{(-1)}(a)$, $\gamma^{(0)}(a)$ or $\gamma_{g_0}^{(-1)}(a), \gamma_{g_0}^{(0)}(a)$. □

Next, we discuss some properties of the indices and indices tuples.

Theorem 4.16 Let g be a primitive root modulo m and $a \in \mathbb{Z}_m^*$. Then $g^h \equiv a \pmod{m}$ if and only if $h \equiv \gamma_{m,g}(a) \pmod{\varphi(m)}$.

Theorem 4.17 Let g be a primitive root modulo m and $a, b \in \mathbb{Z}_m^*$. Then

$$\gamma_{m,g}(ab) \equiv \gamma_{m,g}(a) + \gamma_{m,g}(b) \pmod{\varphi(m)}.$$

Theorem 4.18 Let g, g' be two distinct primitive roots modulo m and $a \in \mathbb{Z}_m^*$. Then

$$\gamma_{m,g'}(a) \equiv \gamma_{m,g'}(g) \cdot \gamma_{m,g}(a) \pmod{\varphi(m)}.$$

This theorem is similar to the change base formula for logarithms. The proofs of the above three theorems are fairly simple, and they are left for the reader. We now state several more theorems that concern the relations between exponents and indices.

Theorem 4.19 Let g be a primitive root modulo m and $(a, m) = 1$. Then

$$\delta_m(a) = \frac{\varphi(m)}{(\gamma_{m,g}(a), \varphi(m))}.$$

Furthermore, when there exists a primitive root modulo m, then for each positive integer $d|\phi(m)$, there are exactly $\phi(d)$ elements in the reduced system of residues modulo m having exponent d. In particular, there are exactly $\phi(\phi(m))$ primitive roots.

Proof By property 6 from Section 4.1, we know that

$$\delta_m(g^k) = \frac{\delta_m(g)}{(k, \delta_m(g))}.$$

Let $a = g^k$, then since $\delta_m(g) = \phi(m)$ and $k = \gamma_{m,g}(a)$, holds.

As g is a primitive root modulo m, $g^0 = 1, g^1, \cdots, g^{\phi(m)-1}$ form a reduced system of residues modulo m, so for each g^i, $\delta_m(g^i) = d$ if and only if

$$\varphi(m), i = \frac{\varphi(m)}{d}, \ 0 \leq i < \varphi(m).$$

Write $i = t \cdot \varphi(m)/d$, then the above is equivalent to $(d, t) = 1, 0 \leq t < d$. The number of such t is $\varphi(d)$. The theorem is proved.

The proof of Theorem 4.19 tells us that the $\varphi(\varphi(m))$ primitive roots are g^t, $0 \leq t < \varphi(m)$, $(t, \varphi(m)) = 1$.

We can list the result of the smallest absolute residue of $g^i, 1 \leq i \leq \varphi(m)$ modulo m according to the order of the indices or the reduced system of residues. Such a list is called an indices table.

Example 4.20 *Construct the indices table of the primitive root 3 modulo 17.*

$\gamma_{17,3}(a)$	0	1	2	3	4	5	6	7	8	9	10	11	12	13	14	15
a	1	3	9	10	13	5	15	11	16	14	8	7	4	12	2	6
$\delta(a)$	1	16	8	16	4	16	8	16	2	16	8	16	4	16	8	16

Theorem 4.21 *Given modulus 2^α, if $a \equiv (-1)^j 5^h \pmod{2^\alpha}$, then*

$$j \equiv \gamma^{(-1)}(a) \equiv (a-1)/2 \pmod 2$$

and

$$h \equiv \gamma^{(0)}(a) \pmod{2^{\alpha-2}}.$$

Theorem 4.22 *Given modulus 2^α, if $(ab, 2) = 1$, then*

$$\gamma^{(-1)}(ab) \equiv \gamma^{(-1)}(a) + \gamma^{(-1)}(b) \pmod 2$$

and

$$\gamma^{(0)}(ab) \equiv \gamma^{(0)}(a) + \gamma^{(0)}(b) \pmod{2^{\alpha-2}}.$$

Proofs of these two theorems are fairly simple and are left as exercises for the reader.

Theorem 4.23 *Let* $(a, 2) = 1$, *then*

$$
\delta_{2^\alpha}(a) = \begin{cases} \dfrac{2^{\alpha-2}}{(\gamma^{(0)}(a), 2^{\alpha-2})}, & 0 < \gamma^{(0)}(a) < 2^{\alpha-2}; \\[3mm] \dfrac{2}{(\gamma^{(-1)}(a), 2)}, & \gamma^{(0)}(a) = 0. \end{cases}
$$

Proof As $\gamma^{(0)}(a) = 0$ if and only if $a \equiv (-1)^{\gamma^{(-1)}(a)} \equiv \pm 1 \pmod{2^\alpha}$, it is easy to see that the theorem holds for this case. If $0 < \gamma^{(0)}(a) < 2^{\alpha-2}$, then we must have $a \not\equiv 1 \pmod{2^\alpha}$, so $2 | \delta_{2^\alpha}(a)$. Let $b = 5^{\gamma^{(0)}(a)}$ and denote $\delta(a) = \delta_{2^\alpha}(a)$, $\delta(b) = \delta_{2^\alpha}(b)$, then from the properties of an index,

$$
\delta(b) = \frac{2^{\alpha-2}}{(\gamma^{(0)}(a), 2^{\alpha-2})}.
$$

Since $0 < \gamma^{(0)}(a) < 2^{\alpha-2}$, we have $2 | \delta(b)$. The fact $2 | \delta(a)$ yields

$$
1 \equiv a^{\delta(a)} \equiv ((-1)^{\gamma^{(-1)}(a)} b)^{\delta(a)} \equiv b^{\delta(a)} \pmod{2^\alpha}.
$$

The fact $2 | \delta(b)$ yields

$$
a^{\delta(b)} \equiv ((-1)^{\gamma^{(-1)}(a)} b)^{\delta(b)} \equiv b^{\delta(b)} \equiv 1 \pmod{2^\alpha}.
$$

Therefore, $\delta(a) | \delta(b)$ and $\delta(b) | \delta(a)$. That is, $\delta(a) = \delta(b)$ and the result holds.

Remark Let $d = 2^j$, $1 < j \le 2^{\alpha-2}$, then $\delta_{2^\alpha}(a) = d = 2^j$ if and only if

$$
(\gamma^{(0)}(a), 2^{\alpha-2}) = 2^{\alpha-2-j}, 0 < \gamma^{(0)}(a) < 2^{\alpha-2}.
$$

Example 4.24 *Construct the indices table for bases -1 and 5 modulo 2^5.*

$\gamma^{-1}(a)$	0	0	0	0	0	0	0	0	1	1	1	1	1	1	1	1
$\gamma^0(a)$	0	1	2	3	4	5	6	7	0	1	2	3	4	5	6	7
a	1	5	25	29	17	21	9	13	31	27	7	3	15	11	23	19
$\delta(a)$	1	8	4	8	2	8	4	8	2	8	4	8	2	8	4	8

We have just discussed the reduced systems of residues modulo p^α and 2^α; we will begin the construction of reduced systems of residues modulo a general m.

Theorem 4.25 *Let* $m = 2^{\alpha_0} p_1^{\alpha_1} p_2^{\alpha_2} \cdots p_r^{\alpha_r}$, *where* $\alpha_i \geq 1 \ (1 \leq i \leq r)$, $p_j \ (1 \leq j \leq r)$ *are mutually distinct odd primes, and for each* $1 \leq j \leq r$, g_j *is a primitive root of* $p_j^{\alpha_j}$, *then*

$$M_0 M_0^{(-1)} (-1)^{\gamma^{(-1)}} 5^{\gamma^{(0)}} + M_1 M_1^{(-1)} g_1^{\gamma^{(1)}} + \cdots + M_r M_r^{(-1)} g_r^{\gamma^{(r)}},$$
$$0 \leq \gamma^{(j)} < c_j, -1 \leq j \leq r$$

form a reduced system of residues modulo m, *where*

$$c_{-1} = c_{-1}(\alpha_0) = \begin{cases} 1, & \alpha_0 = 1, \\ 2, & \alpha_0 \geq 2; \end{cases} \quad c_0 = c_0(\alpha_0) = \begin{cases} 1, & \alpha_0 = 1, \\ 2^{\alpha_0 - 2}, & \alpha_0 \geq 2; \end{cases}$$
$$c_j = \varphi(p_j^{\alpha_j}), \ 1 \leq j \leq r, \ m = M_0 2^{\alpha_0} = M_j p_j^{\alpha_j},$$
$$M_j^{-1} M_j \equiv 1 \pmod{p_j^{\alpha_j}}, \ 1 \leq j \leq r.$$

This theorem can be proved easily by using the concepts of index and index tuple together with the Chinese remainder theorem.

Now we define the concept of a general index tuple.

Definition 4.26 For any given a with $(a, m) = 1$, there must be a unique tuple $\gamma^{(j)} = \gamma^{(j)}(a) \ (-1 \leq j \leq r)$ that satisfies the above theorem, such that

$$a \equiv M_0 M_0^{(-1)} (-1)^{\gamma^{(-1)}} 5^{\gamma^{(0)}} + M_1 M_1^{(-1)} g_1^{\gamma^{(1)}} + \cdots$$
$$+ M_r M_r^{(-1)} g_r^{\gamma^{(r)}} \pmod{m}.$$

We call $\gamma^{(-1)}(a), \gamma^{(0)}(a); \gamma^{(1)}(a), \cdots, \gamma^{(r)}(a)$ the index tuple of a modulo m with the base $-1, 5; g_1, \cdots, g_r$. It is denoted as

$$\gamma_m(a) = \{\gamma^{(-1)}(a), \gamma^{(0)}(a); \gamma^{(1)}(a), \cdots, \gamma^{(r)}(a)\}. \qquad \square$$

Example 4.27 *Find the reduced system modulo* $m = 2^3 \times 5^2 \times 7^2 \times 11^2$.

Solution: Let

$$
\begin{array}{llll}
M_0 = 5^2 \times 7^2 \times 11^2 \equiv 1 & \pmod{2^3}, & M_0^{-1} \equiv 1 & \pmod{2^3}, \\
M_1 = 2^3 \times 7^2 \times 11^2 \equiv 7 & \pmod{5^2}, & M_1^{-1} \equiv -7 & \pmod{5^2}, \\
M_2 = 2^3 \times 5^2 \times 11^2 \equiv -6 & \pmod{7^2}, & M_2^{-1} \equiv 8 & \pmod{7^2}, \\
M_3 = 2^3 \times 5^2 \times 7^2 \equiv -1 & \pmod{11^2}, & M_3^{-1} \equiv -1 & \pmod{11^2}.
\end{array}
$$

We can verify that $2, 3, 2$ are primitive roots of $5^2, 7^2, 11^2$, respectively. Therefore,

$$x = 5^2 \times 7^2 \times 11^2 \times (-1)^{\gamma^{(-1)}} \times 5^{\gamma^{(0)}} + 2^3 \times 7^2 \times 11^2 \times (-7) \times 2^{\gamma^{(1)}}$$
$$+ 2^3 \times 5^2 \times 11^2 \times 8 \times 3^{\gamma^{(2)}} + 2^3 \times 5^2 \times 7^2 \times (-1) \times 2^{\gamma^{(3)}}$$
$$(\mathrm{mod}\ 2^3 \times 5^2 \times 7^2 \times 11^2),$$

with $0 \le \gamma^{(-1)} < 2$, $0 \le \gamma^{(0)} < 2$, $0 \le \gamma^{(1)} < 20$, $0 \le \gamma^{(2)} < 42$, $0 \le \gamma^{(3)} < 110$ form a reduced system of residues modulo $m = 2^3 \times 5^2 \times 7^2 \times 11^2$.

4.4 NTH POWER RESIDUES

We mentioned the concepts of quadratic residues and nth power residues in Chapter 3. In this section, we will get into them briefly.

Definition 4.28 Let $m \ge 2$, $(a, m) = 1$, $n \ge 2$. If the congruence equation

$$x^n \equiv a \pmod{m} \tag{4.1}$$

has a solution, then a is called an nth power residue modulo m; if the equation has no solution, then a is called an nth power nonresidue modulo m. □

Theorem 4.29 *Let $m \ge 2$, $(a, m) = 1$. If there is a primitive root g modulo m, then the congruence Equation 4.1 has a solution if and only if*

$$(n, \varphi(m)) | \gamma(a),$$

where $\gamma(a) = \gamma_{m,g}(a)$ is the index of a modulo m with the base g. Furthermore, if Equation 4.1 has a solution, then it has exactly $(n, \varphi(m))$ solutions.

Proof If $x_1 \pmod{m}$ is a solution of Equation 4.1, then from $(a, m) = 1$ we get $(x_1, m) = 1$, so there must be y_1 such that

$$x_1 \equiv g^{y_1} \pmod{m}, \quad g^{ny_1} \equiv a \pmod{m},$$

and hence,

$$ny_1 \equiv \gamma(a) \pmod{\varphi(m)}$$

by the property of index. This shows that $y \equiv y_1 \pmod{\varphi(m)}$ is a solution of linear congruence equation

$$ny \equiv \gamma(a) \pmod{\varphi(m)}.$$

This implies $(n, \varphi(m))|\gamma(a)$. As the deduction process can be reversed, the necessity is immediate.

Theorem 4.29 gives a concrete method for solving Equation 4.1 theoretically, when a primitive root modulo m exists.

1. Find the index $\gamma(a)$ of a from the index table.

2. Solve the congruence equation $ny \equiv \gamma(a) \pmod{\varphi(m)}$.

3. If $ny \equiv \gamma(a) \pmod{\varphi(m)}$ has a solution, then for each solution $y_1 \pmod{\varphi(m)}$, find x_1 such that $x_1 \equiv g^{y_1} \pmod{m}$ using the index table. All such $x_1 \pmod{m}$ form the set of all solutions of $x^n \equiv a \pmod{m}, n \geq 2$.

Remark The reasons that we say that the above method is only a theoretical one are as follows:

1. When the modulus m is a composite, finding $\varphi(m)$ is equivalent to integer factorization. This can be done in theory, but when it comes to the practical implementation, there is no efficient algorithm to compute $\varphi(m)$ at the moment.

2. Even if $\varphi(m)$ is given, finding $\gamma(a)$ is to solve the discrete logarithm problem which is again a difficult problem.

Remark When $\varphi(m)$ is given and $(\varphi(m), n) = 1$, there is a simple method to solve $x^n \equiv a \pmod{m}$. In fact,

$$x = x^{nn^{-1}(\bmod \varphi(m))} = a^{n^{-1}(\bmod \varphi(m))} \pmod{m}$$

is a solution.

Example 4.30 *Solve the congruence equation $x^{10} \equiv 13 \pmod{17}$.*

Solution 1: It is easy to see that 3 is a primitive root modulo 17 and $\gamma_{17,3}(13) = 4$, the congruence equation

$$10y \equiv 4 \pmod{16}$$

has solutions $y \equiv 2, -6 \pmod{16}$. We get two solutions $8, 9 \pmod{17}$ of the equation by simple calculation.

Solution 2: First, we get the square roots ± 8 of 13 modulo 17, then we compute 5^{-1} (mod 16) $= 13$. Therefore,

$$x \equiv x^{5 \times 5^{-1}} \equiv (\pm 8)^{13} \equiv \pm 8 \equiv 8, 9 \quad (\text{mod } 17).$$

The nth power residues modulo m has the following property.

Property 1 If there is a primitive root modulo m and $n \geq 2$, then there are exactly $\varphi(m)/(n, \varphi(m))$ nth power residues modulo m in a reduced system of residues modulo m.

Theorem 4.31 *Let $\alpha \geq 3$ and a be an odd number. Assume that the index tuple of a modulo 2^α with the base $-1, 5$ is $\gamma^{(-1)}, \gamma^{(0)}$. Then a is an nth power residues modulo 2^α if and only if*

$$(n, 2) | \gamma^{(-1)}(a), \quad (n, 2^{\alpha - 2}) | \gamma^{(0)}(a).$$

If a is an nth power residues modulo 2^α, then $x^n \equiv a$ (mod 2^α) has exactly $(n, 2) \cdot (n, 2^{\alpha - 2})$ solutions. In other words, when n is an odd number, there is exactly one solution; when n is an even number, there are exactly $2 \cdot (n, 2^{\alpha - 2})$ solutions.

Proof Since a is an odd number, the solutions x of Equation 4.1 can only be taken from a reduced system of residues modulo 2^α. Therefore, we can assume

$$x = (-1)^u 5^v, \ 0 \leq u < 2, \ 0 \leq v < 2^{\alpha - 2}.$$

This turns Equation 4.1 to a congruence equation with two variables:

$$\begin{cases} (-1)^{nu} 5^{nv} \equiv (-1)^{\gamma^{(-1)}(a)} 5^{\gamma^{(0)}(a)} \quad (\text{mod } 2^\alpha), \\ 0 \leq u < 2, 0 \leq v < 2^{\alpha - 2}. \end{cases}$$

From the properties of exponents, this is equivalent to the following system of congruence equations:

$$\begin{cases} nu \equiv \gamma^{(-1)}(a) \quad (\text{mod } 2), & 0 \leq u < 2; \\ nv \equiv \gamma^{(0)}(a) \quad (\text{mod } 2^{\alpha - 2}), & 0 \leq v < 2^{\alpha - 2}. \end{cases}$$

From the theory of congruence equations, the first equation (note that u is taken from a complete system of residues modulo 2) has a solution if and only if

$$(n, 2)|\gamma^{(-1)}(a).$$

If it has a solution, then it has exactly $(n, 2)$ solutions. The second equation (note that v is taken from a complete system of residues modulo $2^{\alpha-2}$) has a solution if and only if

$$(n, 2^{\alpha-2})|\gamma^{(0)}(a).$$

If it has a solution, then it has exactly $(n, 2^{\alpha-2})$ solutions. The theorem is proved.

Example 4.32 *Solve congruence equation* $x^7 \equiv 29 \pmod{2^5}$.

Solution: It is easy to see that the index tuple of 29 is $\gamma^{(-1)}(29) = 0$, $\gamma^{(0)}(29) = 3$. Thus, the problem is reduced to solve two linear congruence equations:

$$\begin{cases} 7u \equiv 0 \pmod{2}, \\ 7v \equiv 3 \pmod{2^3}. \end{cases}$$

This gives

$$\begin{cases} u \equiv 0 \pmod{2}, \\ v \equiv 5 \pmod{2^3}, \end{cases}$$

and $x \equiv (-1)^0 5^5 \pmod{2^5}$ is the solution to the equation.

EXERCISES

4.1 Let $(m_1, m_2) = 1$. Prove that for any a_1, a_2, there must be an a such that

$$\delta_{m_1 m_2}(a) = [\delta_{m_1}(a_1), \delta_{m_2}(a_2)].$$

4.2 Construct exponent tables for $m = 5, 11, 13, 15, 19, 20$.

4.3 Find $\delta_{3 \times 17}(10)$, $\delta_{11^2}(2)$.

4.4 Let $m = 2^\alpha, \alpha \geq 3$. Prove that $\delta_m(a) = 2^{\alpha-2}$ if and only if $a \equiv \pm 3 \pmod 8$.

4.5 Let prime $p > 2$ and $p - 1$ has standard prime factorization $q_1^{\beta_1} \cdots q_r^{\beta_r}$. Prove that

(1) For any j $(1 \leq j \leq r)$, there is an a_j whose exponent modulo p is $q_j^{\beta_j}$ (you should not use primitive roots modulo p).

(2) $a_1 \cdots a_r$ is a primitive root modulo p.

4.6 Let $m = 2^{\alpha_0} p_1^{\alpha_1} p_2^{\alpha_2} \cdots p_r^{\alpha_r}$, where p_j are distinct odd primes and $(m, a) = 1$. Prove that $a^{\lambda(m)} \equiv 1 \pmod{m}$, where

$$\lambda(m) = [2^{c_0}, \varphi(p_1^{\alpha_1}), \cdots, \varphi(p_r^{\alpha_r})], \quad c_0 = \begin{cases} 0, & \alpha_0 = 0, 1; \\ 1, & \alpha_0 = 2; \\ \alpha_0 - 2, & \alpha_0 \geq 3. \end{cases}$$

4.7 Let p be odd prime and $q_1, q_2, \cdots, q_s$ are distinct prime factors of $p - 1$. Prove that g is a primitive root modulo p if and only if $g^{p-1/q_j} \not\equiv 1 \pmod{p}, j = 1, 2 \cdots s$.

4.8 Find primitive roots for modulus $23, 29, 41, 53, 67$, and 73.

4.9 For each $p = 5, 7, 11, 13, 17$, find a primitive root g modulo p such that g is not a primitive root modulo p^2.

4.10 Find the smallest prime modulo which 11 is a primitive root.

4.11 If $p = 2^{2^k} + 1$ is a prime, find a necessary and sufficient condition under which 7 is a primitive root modulo p.

4.12 Find a reduced system of residues modulo $m = 3 \times 13 \times 23 \times 43$.

4.13 Solve congruence equations:

(1) $x^5 \equiv 3 \pmod{61}$.

(2) $6x^{10} \equiv 5 \pmod{17}$.

4.14 For which values of a, does $ax^8 \equiv 5 \pmod{17}$ have a solution?

4.15 Let prime $p > 2$. Prove that the congruence equation $x^4 \equiv -1 \pmod{p}$ has solution if and only if $p \equiv 1 \pmod{8}$.

4.16 Let p be a prime and $2 \nmid \delta_p(a)$ prove that the congruence equation $a^x + 1 \equiv 0 \pmod{p}$ has no solution.

4.17 Let prime $p \equiv 3 \pmod 4$. Prove that a is a 4th power of residue modulo p if and only if $\left(\dfrac{a}{p} \right) = 1$.

CHAPTER 5

Some Elementary Results for Prime Distribution

T HE STUDY of various properties of prime numbers is not only the most important content in number theory, but it also plays an important role in cryptography. In previous chapters, we have had discussions of some of the properties of prime numbers. This chapter will explore some other properties and the problem about how primes are distributed in the set of positive integers by describing the prime number theorem. We also introduce an important result for the prime distribution in an arithmetic progression—the prime number theorem in arithmetic progressions. The results of this chapter have many applications in cryptography. For example, the general prime number theorem and the prime number theorem in an arithmetic progression explain the average probabilities of prime distribution in the set of positive integers and in an arithmetic progression, respectively. These provide the theoretical basis for probabilistic algorithms of generating prime numbers.

5.1 INTRODUCTION TO THE BASIC PROPERTIES OF PRIMES AND THE MAIN RESULTS OF PRIME NUMBER DISTRIBUTION

In this section, we briefly introduce several properties about prime numbers and two important results of prime number distribution—the prime number theorem and prime number theorem in arithmetic progressions. We will recap some properties of prime numbers covered in previous chapters.

Property 1 (The Fundamental Theorem of Arithmetic) Every integer $n > 1$ has the following prime factorization:

$$n = p_1^{\alpha_1} p_2^{\alpha_2} \cdots p_r^{\alpha_r},$$

where $p_1, \cdots, p_r$ are distinct prime factors. This factorization is unique except for the order of the factors.

Property 1 explains a relationship between integers and prime numbers; that is, a positive integer can be expressed as a product of prime numbers. Factorization of big integers is a difficult problem. In public key cryptography, a large number of cryptographic algorithms are based on the problem of integer factorization.

Another description of the fundamental theorem of arithmetic is in an analytic equivalent form.

Theorem 5.1 *The fundamental theorem of arithmetic is equivalent to*

$$\prod_p \left(1 - \frac{1}{p^s}\right)^{-1} = \sum_{n=1}^{\infty} \frac{1}{n^s}, \quad s > 1. \tag{5.1}$$

Equation 5.1 is the famous Euler's product formula.

For the distribution of primes, we have proved in Chapter 1 that there are infinitely many primes.

Property 2 There are infinitely many primes, that is,

$$\lim_{x \to \infty} \pi(x) = \infty,$$

where $\pi(x)$ denotes the number of primes that are less than or equal to x.

There exists a more accurate estimation for the main term of $\pi(x)$, that is the famous prime number theorem, or the prime number theorem without error term.

Theorem 5.2 (*Prime number theorem*)

$$\pi(x) \sim \frac{x}{\ln x}, \quad x \to \infty. \tag{5.2}$$

For the prime number theorem, Legendre suggested the following accurate estimation as early as 1800:

$$\pi(x) \sim \frac{x}{\ln x - 1.08366}, \ x \to \infty.$$

It is easy to prove that $\int_2^x \frac{du}{\ln u} \sim \frac{x}{\ln x}$. Denoting $\mathrm{Li}(x) = \int_2^x \frac{du}{\ln u}$, then the prime number theorem can also be described as

$$\pi(x) \sim \mathrm{Li}(x), \ x \to \infty.$$

This was the conjecture in the integral form given by Gauss.

However, the conjecture of Legendre and Gauss remained unproven until 1896 when Hadamard and de la Vallée Poussin provided independent proofs by using some deep theory from complex analysis. No elementary proof of this result was available until 1949, when Selberg and Erdös independently found such a proof. Selberg was awarded the Fields medal for his proof. We will provide proof to a weaker version of the prime number theorem—the Chebyshev inequality.

Let $R(x) = \pi(x) - \mathrm{Li}(x)$, the prime number theorem is in fact equivalent to the statement that $R(x) = o(\frac{x}{\ln x})$. The prime number theorem with an error term means a more precise estimation of $R(x)$. The estimation of the prime number theorem with an error term is an important and interesting topic of analytic number theory. There have been many results on this topic, but the proofs involve some deeper knowledge from analytic number theory; interested readers are referred to the literature [6]. Here we just state a result proved by Balog in 1981.

Theorem 5.3 (*Balog*) $\pi(x) = \mathrm{Li}(x) + O(x(\ln x)^{-\frac{5}{4}}(\ln \ln x))$.

Finally, let us briefly describe the problem of prime distribution in arithmetic progressions.

In cryptography, we often need to choose large primes with some special forms. The most common case is to choose prime p such that the prime factors of $p - 1$ are known. This kind of prime is usually taken from some arithmetic progressions.

Consider the following arithmetic progression:

$$b, b+a, b+2a, \cdots, b+ka, \cdots, \tag{5.3}$$

where $a \geq 3, 1 \leq b < a, (a, b) = 1$.

We already know that there are infinitely many primes among natural numbers. Are there also infinitely many primes in the above arithmetic progression? Euler stated that the arithmetic progression (5.3) contains infinitely many primes when $b = 1$; later, Legendre clearly conjectured that equation (5.3) contains infinitely many primes. However, neither gave a proof. Although the result was proved for many special cases of a and b, it remained a very difficult problem for the general case. In 1837, Dirichlet proved the conjecture for the case that a is a prime. He later proved the general case by using his class number formula for quadratic forms. We will not include a proof here and ask interested readers to read related materials.

The problem of prime distribution has always been one of the central problems in the study of number theory. From our previous discussion about the prime number theorem, we have gotten the estimation of the main term of the number $\pi(x)$ of primes that are less than or equal to x. Is it possible to estimate the number of primes that are less than or equal to x within the arithmetic progression

$$\{ak + b | k = 0, 1, 2, \cdots\},$$

where $a \geq 3, 1 \leq b < a, (a, b) = 1$? The answer is yes. However, its proof requires deep knowledge from analytic number theory, which is beyond the scope of elementary number theory. Interested readers are referred to relevant references.

In this section, we only state a result about the prime distribution of arithmetic progressions.

Definition 5.4 Let $\pi(x; a, b)$ denote the number of primes that are less than or equal to x in the arithmetic progression

$$\{ak + b | k = 0, 1, 2, \cdots\},$$

where $a \geq 3, 1 \leq b < a, (a, b) = 1$, that is,

$$\pi(x; a, b) = \{p | p \text{ is a prime with } p \equiv b \pmod{a}, \text{ and } p < x\}. \quad \square$$

For an estimation of $\pi(x; a, b)$, we have the following.

Theorem 5.5 *If $(a, b) = 1$, then*

$$\pi(x; a, b) = \frac{1}{\varphi(a)} \operatorname{Li}(x) + O(xe^{-c\sqrt{\log x}}),$$

where $a \leq (\log x)^A$ with some constant $A > 0$, and $c > 0$ is a constant depending on A. In particular,

$$\pi(x) = \mathrm{Li}(x) + O(xe^{-c\sqrt{\log x}}).$$

By Theorem 5.5 and $\pi(x) \sim \dfrac{x}{\ln x}$, we know that the average probability of the prime distribution in arithmetic progression $\{ak + b | k = 0, 1, 2, \cdots\}$ is $\dfrac{1}{\varphi(a)\ln x}$. This provides a theoretical basis for a probabilistic polynomial-time algorithm to find primes of the form $p \equiv b$ (mod a).

5.2   PROOF OF THE EULER PRODUCT FORMULA

We gave an analytic equivalent form of the fundamental theorem of arithmetic in Section 5.1—Euler's product formula. We present a theoretical proof of Euler's product formula in this section. First, we need to prove two related lemmas.

Lemma 5.6 *For real number $s > 1$, the infinite product*

$$\prod_{p} \left(1 - \frac{1}{p^s}\right)^{-1} \tag{5.4}$$

is convergent to a number that is greater than 1, where the product is taken over all primes.

Proof   From the property of logarithm, we have

$$0 < \frac{1}{p^s} < \ln\left(1 - \frac{1}{p^s}\right)^{-1}. \tag{5.5}$$

Therefore, when $s > 1$,

$$\sum_{p} \frac{1}{p^s} < \sum_{p} \ln\left(1 - \frac{1}{p^s}\right)^{-1} < \sum_{p} \frac{1}{p^s - 1} < 2\sum_{p} \frac{1}{p^s} < 2\sum_{n=1}^{\infty} \frac{1}{n^s},$$

where the summation $\sum_p$ runs over all primes p. Since the series $\sum_{n=1}^{\infty} n^{-s}$ converges when $s > 1$, so is the positive series

$$\sum_{p} \ln\left(1 - \frac{1}{p^s}\right)^{-1}.$$

This implies that the infinite product in Equation 5.1 is convergent. It is obvious that its value is greater than 1. The proof is complete.

Without assuming the fundamental theorem of arithmetic, let $c(n)$ be the number of representations of n in the following form: $n = p_1^{\alpha_1} p_2^{\alpha_2} \cdots p_r^{\alpha_r}$, where $p_1, \cdots, p_r$ are distinct primes, and the representations are up to ordering of the factors. Then the following holds.

Lemma 5.7

$$\prod_p \left(1 - \frac{1}{p^s}\right)^{-1} = \sum_{n=1}^{\infty} \frac{c(n)}{n^s}. \qquad (5.6)$$

Proof For real number $s > 1$,

$$\left(1 - \frac{1}{p^s}\right)^{-1} = 1 + \frac{1}{p^s} + \frac{1}{p^{2s}} + \cdots.$$

For any given positive integer N, let k be such that $2^{k-1} \le N < 2^k$, then

$$\sum_{n=1}^{N} \frac{c(n)}{n^s} \le \prod_{p \le N} \left(1 + \frac{1}{p^s} + \frac{1}{p^{2s}} + \cdots + \frac{1}{p^{ks}}\right) \le \prod_p \left(1 - \frac{1}{p^s}\right)^{-1}, \qquad (5.7)$$

where $s > 1$. By Lemma 5.6, the above infinite product converges, so the positive series of the right-hand side of Lemma 5.6 converges and

$$\sum_{n=1}^{\infty} \frac{c(n)}{n^s} \le \prod_p \left(1 - \frac{1}{p^s}\right)^{-1}. \qquad (5.8)$$

Conversely, since

$$\prod_{p \le M} \left(1 + \frac{1}{p^s} + \frac{1}{p^{2s}} + \cdots + \frac{1}{p^{hs}}\right) \le \sum_{n=1}^{\infty} \frac{c(n)}{n^s}, s > 1,$$

we get by letting $h \to +\infty$,

$$\prod_{p \le M} \left(1 - \frac{1}{p^s}\right)^{-1} \le \sum_{n=1}^{\infty} \frac{c(n)}{n^s}, s > 1.$$

The lemma is proved because of this and (5.7).

The proof of Theorem 5.1 Assume the fundamental theorem of arithmetic; then we have $c(n) = 1$, and hence, Theorem 5.1 holds.

Conversely, assuming (5.1), then by Lemma 5.7,

$$\sum_{n=1}^{\infty} \frac{c(n)-1}{n^s} = 0, s > 1.$$

Since $c(n) - 1 \geq 0$ for all n, so we must have $c(n) = 1$, and hence, the fundamental theorem of arithmetic is true.

5.3 PROOF OF A WEAKER VERSION OF THE PRIME NUMBER THEOREM

In this section, we give a proof of Chebyshev inequalities—a weaker version of the prime number theorem.

We first define the Möbius function $\mu(n)$ as

$$\mu(n) = \begin{cases} 1, & n = 1, \\ (-1)^s, & n = p_1 p_2 \cdots p_s, p_1 < \cdots < p_s, \\ 0, & \text{otherwise,} \end{cases}$$

and $\mu(n)$ has the following important properties.

Lemma 5.8

$$\sum_{d|n} \mu(d) = \begin{cases} 1, & n = 1, \\ 0, & n > 1. \end{cases} \tag{5.9}$$

Lemma 5.9 *Let $x > 0$ and $P_s = p_1, p_2, \cdots, p_s$, where $p_1, p_2, \cdots, p_s$ are the first s primes; let $\varphi(x, s)$ denote the number of positive integers that are less than or equal to x and are not divisible by p_i ($1 \leq i \leq s$). Then,*

$$\varphi(x, s) = \sum_{d|P_s} \mu(d) \left[\frac{x}{d}\right]. \tag{5.10}$$

Proof From (5.9), we get

$$\varphi(x, s) = \sum_{n \leq x} \sum_{d|(n, P_s)} \mu(d) = \sum_{d|P_s} \mu(d) \sum_{n \leq x, d|n} 1 = \sum_{d|P_s} \mu(d) \left[\frac{x}{d}\right].$$

The proof is complete.

Lemma 5.10 *Let s be a positive integer and $x > s$, then*

$$\pi(x) < x \prod_{i=1}^{s} \left(1 - \frac{1}{p_i}\right) + 2^{s+1}, \tag{5.11}$$

where $p_1, p_2, \cdots, p_s$ are the first s primes.

Proof Since the prime numbers that are greater than p_s and less than or equal to x cannot be divisible by the first s primes, so

$$\pi(x) \leq s + \varphi(x, s).$$

By Lemma 5.9, we get

$$\pi(x) \leq s + \sum_{d \mid P_s} \mu(d) \left[\frac{x}{d}\right]$$

$$= s + \left(x - \sum_{i=1}^{s} \left[\frac{x}{p_i}\right] + \sum_{1 \leq i \leq j \leq s} \left[\frac{x}{p_i p_j}\right] + \cdots + (-1)^s \left[\frac{x}{p_1 p_2 \cdots p_s}\right]\right)$$

$$< s + x \left(1 - \sum_{i=1}^{s} \frac{1}{p_i} + \cdots + (-1)^s \left(\frac{1}{p_1 p_2 \cdots p_s}\right)\right)$$

$$+ \left(\sum_{i=1}^{s} 1 + \sum_{1 \leq i \leq j \leq s} 1 + \cdots + 1\right)$$

$$< s + x \prod_{i=1}^{s} \left(1 - \frac{1}{p_i}\right) + \left(1 + \binom{s}{1} + \binom{s}{2} + \cdots + 1\right)$$

$$= x \prod_{i=1}^{s} \left(1 - \frac{1}{p_i}\right) + s + (1+1)^s.$$

This immediately yields (5.11). The proof is complete.

Lemma 5.11

$$\prod_{p} \left(1 - \frac{1}{p}\right) = 0. \tag{5.12}$$

Proof Let N be a sufficiently large positive integer. Then it is obvious that

$$\prod_{p}\left(1-\frac{1}{p}\right)^{-1} > \prod_{p\leq N}\left(1-\frac{1}{p}\right)^{-1} = \prod_{p\leq N}\left(\sum_{r=0}^{\infty}\frac{1}{p^r}\right) > \sum_{n=1}^{N}\frac{1}{n}.$$

From

$$\lim_{N\to\infty}\sum_{n\leq N}\frac{1}{n} = \infty,$$

it is immediate that

$$\prod_{p}\left(1-\frac{1}{p}\right) = 0.$$

The proof is complete.

We are now able to prove Theorem 5.12.

Theorem 5.12 $\lim\limits_{x\to\infty}\dfrac{\pi(x)}{x} = 0.$

Proof By Lemma 5.10,

$$\pi(x) < x\prod_{i=1}^{s}\left(1-\frac{1}{p_i}\right) + 2^{s+1}.$$

Set $s = \left[\dfrac{\ln x}{2\ln 2}\right] - 1$, then

$$0 < \frac{\pi(x)}{x} < \prod_{i=1}^{\left[\frac{\ln x}{2\ln 2}\right]-1}\left(1-\frac{1}{p_i}\right) + \frac{2^{\frac{\ln x}{2\ln 2}}}{x}.$$

The right-hand side of the above goes to zero as $x \to \infty$, so

$$\lim_{x\to\infty}\frac{\pi(x)}{x} = 0.$$

Or $\pi(x) = o(x)$, $x \to \infty$.

Theorem 5.12 indicates that the "density" of primes in the set of natural numbers is very small (with probability 0).

Next, we present better estimations of the upper and lower bounds of the prime distribution function $\pi(x)$—Chebyshev inequalities.

Theorem 5.13 *Let $x \geq 2$. Then*

$$\left(\frac{\ln 2}{3}\right)\frac{x}{\ln x} < \pi(x) < (6\ln 2)\frac{x}{\ln x}, \tag{5.13}$$

and

$$\left(\frac{1}{6\ln 2}\right)n\ln n < p_n < \left(\frac{8}{\ln 2}\right)n\ln n, \ n \geq 2, \tag{5.14}$$

where p_n denotes the nth prime number.

Proof We first prove (5.13). Let m be a positive integer and set

$$M = \frac{(2m)!}{(m!)^2}.$$

It can be proved that M is a positive integer and

$$\begin{aligned}
\ln M &= \ln(2m) - 2\ln(m!) \\
&= \sum_{p \leq m} \{a(p, 2m) - 2a(p, m)\}\ln p \\
&= \sum_{m < p \leq 2m} a(p, 2m)\ln p,
\end{aligned} \tag{5.15}$$

where

$$a(p, n) = \sum_{j=1}^{\infty}\left[\frac{n}{p^j}\right]. \tag{5.16}$$

It is easy to see that when $m < p \leq 2m$,

$$a(p, 2m) = 1. \tag{5.17}$$

When $p \leq m$, from $0 \leq [2y] - 2[y] \leq 1$ and (5.15), we get

$$\begin{aligned}
0 \leq a(p, 2m) - 2a(p, m) &= \sum_{j=1}^{\infty}\left\{\left[\frac{2m}{p^j}\right] - 2\left[\frac{m}{p^j}\right]\right\} \\
&\leq \sum_{p \leq 2m} 1 = \left[\frac{\ln(2m)}{\ln p}\right].
\end{aligned} \tag{5.18}$$

Therefore, by (5.15), (5.17), and (5.18), we have

$$\sum_{m < p \leq 2m}\ln p \leq \ln m \leq \sum_{p \leq 2m}\left[\frac{\ln(2m)}{\ln p}\right]\ln p. \tag{5.19}$$

So

$$\{\pi(2m) - \pi(m)\} \ln m \le \ln M \le \pi(2m) \ln(2m). \tag{5.20}$$

We can estimate the upper and lower bounds for M directly. We have

$$M = \frac{2m}{m} \cdot \frac{2m-1}{m-1} \cdots \frac{m+1}{1} \ge 2^m, \tag{5.21}$$

and

$$M = \frac{(2m)!}{(m!)^2} < (1+1)^{2m} = 2^{2m}. \tag{5.22}$$

Combining the three relations above, we get

$$\pi(2m) \ln(2m) \ge m \ln 2$$

and

$$\{\pi(2m) - \pi(m)\} \ln m < 2m \ln 2. \tag{5.23}$$

If $x \ge 6$, setting $m = \left[\frac{x}{2}\right] > 2$, then it is obvious that $2m \le x < 3m$. So the left half of (5.13) is established.

When $m = 2^k$, we see that from Equation 5.23

$$k\{\pi(2^{k+1}) - \pi(2^k)\} < 2^{k+1}.$$

Together with the fact that $\pi(2^{k+1}) \le 2^k$ $(k \ge 0)$, this implies

$$(k+1)\pi(2^{k+1}) - k\pi(2^k) < 3 \times 2^k.$$

Summing up the above for k from 0 to $l-1$, we get

$$l \cdot \pi(2^l) < 3 \times 2^l.$$

For any $x \ge 2$, let h be such that $2^{h-1} < x \le 2^h$. Then we have

$$\pi(x) \le \pi(2^k) < 3 \times \frac{2^k}{k} < (6 \ln 2) \frac{x}{\ln x}.$$

This establishes the right half of (5.13).

In the above relation, if we take $x = p_n$, and note $p_n > n$, then

$$p_n > \left(\frac{1}{6 \ln 2}\right) n \ln p_n > \left(\frac{1}{6 \ln 2}\right) n \ln n.$$

This proves the left half of (5.14).

For $n > 1$ and taking $2m = p_n + 1$ in (5.19), we see that

$$n \ln(p_n + 1) \geq \frac{p_n + 1}{2} \ln 2. \tag{5.24}$$

So,

$$\ln(p_n + 1) \leq \ln\left(\frac{2n}{\ln 2}\right) + \ln\ln(p_n + 1). \tag{5.25}$$

Note that when $s > -1$, we have $\ln(1 + s) = \int_0^s \frac{dt}{1+t} \leq s$. By setting $s = y/2 - 1$, we get

$$\ln y \leq \frac{y}{2} - (1 - \ln 2) < \frac{y}{2}, y > 0.$$

Now take $y = \ln(p_n + 1)$, then by (5.25) we have

$$\ln(p_n + 1) \leq 2\ln\left(\frac{2n}{\ln 2}\right) < 4\ln n, n \geq 3.$$

This, together with Equation 5.21, implies the right half of (5.14) for $n \geq 3$. We can simply verify the right half of Equation 5.14 directly for $n < 3$.

Some estimations of the average distribution of primes can be obtained immediately from Theorem 5.13. To this end, we need the following lemma.

Lemma 5.14 *Let $y \geq 2$. Then we have*

$$\ln\ln([y] + 1) - \ln\ln 2 < \sum_{2 \leq k \leq y} \frac{1}{k \ln k} < \ln\ln[y] + \frac{1}{2\ln 2} - \ln\ln 2, \tag{5.26}$$

and

$$[y](\ln[y] - 1) + 1 < \sum_{2 \leq k \leq y} \frac{1}{k \ln k}$$
$$< ([y] + 1)\{(\ln[y] + 1) - 1\} + 2 - 2\ln 2. \tag{5.27}$$

Proof We have

$$\int_k^{k+1} \frac{dt}{t \ln t} < \frac{1}{k \ln k} < \int_{k-1}^k \frac{dt}{t \ln t}, k \geq 3.$$

Therefore,

$$\sum_{2\leq k\leq y} \frac{1}{k\ln k} < \frac{1}{2\ln 2} + \int_2^{[y]} \frac{dt}{t\ln t}$$

$$= \ln\ln[y] + \frac{1}{\ln 2} - \ln\ln 2.$$

$$\sum_{2\leq k\leq y} \frac{1}{k\ln k} > \int_2^{[y]+1} \frac{dt}{t\ln t} = \ln\ln([y]+1) - \ln\ln 2.$$

Equation 5.26 follows immediately from the above two inequalities. Similarly, from

$$\int_{k-1}^{k} \ln t\, dt < \ln k < \int_{k}^{k+1} \ln t\, dt,$$

we get

$$\sum_{1\leq k\leq y} \ln k < \int_2^{[y]+1} \ln t = ([y]+1)\ln([y]+1) - ([y]+1) + 2 - 2\ln 2$$

and

$$\sum_{1\leq k\leq y} \ln k > \int_1^{[y]} \ln t\, dt = [y]\ln[y] - [y] + 1.$$

Thus, Equation 5.27 is proved.

By Lemma 5.14 and Equation 5.14, we have Theorem 5.15.

Theorem 5.15 *For $x \geq 5$, there must exist positive constant $c_1, c_2,$ $\cdots, c_6$ such that*

$$c_1\ln\ln x < \sum_{p\leq x}\frac{1}{p} < c_2\ln\ln x, \tag{5.28}$$

$$c_3 x < \sum_{p\leq x}\ln p < c_4 x, \tag{5.29}$$

$$c_5\ln x < \sum_{p\leq x}\frac{\ln p}{p} < c_6 x. \tag{5.30}$$

Furthermore,

$$\lim_{n\to\infty}\frac{\ln p_n}{\ln n} = 1. \tag{5.31}$$

Proof Equation 5.31 can be obtained from (5.14) immediately. Also by (5.14), it is easy to see that

$$a_1 \ln n < \ln p_n < a_2 \ln n, n \geq 2, \tag{5.32}$$

$$a_3 \ln \ln n < \ln \ln p_n < a_4 \ln \ln n, n \geq 25, \tag{5.33}$$

where a_1, a_2, a_3, a_4 are positive constants independent of n.

We now prove (5.28) through (5.30). Assume $x \geq 10$, and let $p_m \leq x p_{m+1}$. Then $m \geq 2$. We first consider (5.28). From (5.14), we know that there exist positive constants a_5, a_6 such that

$$a_5 \sum_{k=2}^{m} \frac{1}{k \ln k} < \sum_{p \leq x} \frac{1}{p} = \sum_{k=1}^{m} \frac{1}{p_k} < a_6 \sum_{k=2}^{m} \frac{1}{k \ln k} + \frac{1}{2}.$$

Further, using (5.26), $m \geq 25$ implies that there exist positive constants a_7, a_8 such that

$$a_7 \ln \ln(m+1) < \sum_{p \leq x} \frac{1}{p_k} < a_8 \ln \ln m.$$

From (5.33) and $m \geq 25$,

$$\ln \ln m < a_3^{-1} \ln \ln p_m \leq a_3^{-1} \ln \ln x,$$
$$\ln \ln(m+1) > a_4^{-1} \ln \ln p_{m+1} > \ln \ln x.$$

Equation 5.28 follows from the above three relations.

Next, we prove (5.29). From (5.32), we get

$$a_1 \sum_{k=2}^{m} \ln k < \sum_{p \leq x} \ln p = \sum_{k=1}^{m} \ln p_m < a_2 \sum_{k=2}^{m} \ln k + \ln 2.$$

By (5.27) and $m \geq 25$, we know that there exist positive constants a_9, a_{10} such that

$$a_9(m+1)\ln(m+1) < \sum_{p \leq x} \ln p < a_{10} m \ln m,$$

together with (5.27) and $m \geq 25$, we get

$$m \ln m < (6 \ln 2) p_m \leq (6 \ln 2) x,$$

and

$$(m+1)\ln(m+1) > \ln \left(\frac{2}{8}\right) p_{m+1} > \left(\ln \frac{2}{8}\right) x.$$

Equation 5.29 follows from the above three relations.

Finally, we prove (5.30). By (5.14) and (5.32), there are positive constants a_{11}, a_{12} such that

$$\frac{a_{11}}{n} < \frac{\ln p_n}{p_n} < \frac{a_{12}}{n}, \quad n \geq 1.$$

Therefore,

$$a_{11} \sum_{k=1}^{m} \frac{1}{k} < \sum_{p \leq x} \frac{\ln p}{p} = \sum_{k=1}^{m} \frac{\ln p_k}{p_k} < a_{12} \sum_{k=1}^{m} \frac{1}{k}.$$

This, together with

$$\ln(m+1) = \int_1^{m+1} t^{-1} dt < \sum_{k=1}^{m} \frac{1}{k} < 1 + \int_1^m t^{-1} dt = 1 + \ln m,$$

gives us

$$a_{11} \ln(m+1) < \sum_{p \leq x} \frac{\ln p}{p} < 2a_{12} \ln m.$$

By Equation 5.32,

$$\ln m < a_1^{-1} \ln p_m < a_1^{-1} \ln x,$$
$$\ln(m+1) > a_2^{-1} \ln p_{m+1} > a_2^{-1} \ln x.$$

Equation 5.30 follows from the above three relations.

5.4 EQUIVALENT STATEMENTS OF THE PRIME NUMBER THEOREM

To prove the prime number theorem, Chebyshev introduced two important functions to replace $\pi(x)$:

$$\theta(x) = \sum_{p \leq x} \ln p, \tag{5.34}$$

$$\Psi(x) = \sum_{n \leq x} \Lambda(n), \tag{5.35}$$

where $\Lambda(n)$ is defined as

$$\Lambda(n) = \begin{cases} \ln p, & n = p^\alpha, p \text{ is a prime }, \alpha \geq 1, \\ 0, & \text{otherwise.} \end{cases} \tag{5.36}$$

The Λ is usually called the Mangoldt function. Compared to $\pi(x)$, these two functions are more convenient to discuss. We first prove a theorem to explain the relations among these three functions.

Theorem 5.16 *Let $x \geq 2$, then there exists a positive constant c such that*

$$(\ln x - c)\pi(x) < \theta(x) < (\ln(x))\pi(x), \tag{5.37}$$

and

$$\theta(x) \leq \Psi(x) \leq \theta(x) + x^{\frac{1}{2}}\ln x. \tag{5.38}$$

Proof First, let us prove (5.37). We have

$$\theta(x) = \sum_{p \leq x}\ln p = \sum_{k \leq x}\ln k(\pi(k) - \pi(k-1))$$

$$= -\sum_{k=2}^{[x]-1}\pi(k)(\ln(k+1) - \ln k) + \pi([x])\ln[x].$$

Note that $\frac{s}{1+s} \leq \ln(1+s) \leq s$, or equivalently,

$$\frac{1}{y+1} < -\ln\left(1 - \frac{1}{y+1}\right) = \ln\left(1 + \frac{1}{y}\right) < \frac{1}{y}, y \geq 1, \tag{5.39}$$

so we get

$$\pi(x)\ln[x] - \sum_{k=2}^{[x]-1}\frac{1}{\ln k} < \theta(x) < \pi(x)\ln x - \sum_{k=2}^{[x]-1}\frac{1}{k+1}.$$

By Chebyshev inequality, we have

$$\sum_{k=2}^{[x]-1}\frac{\pi(k)}{k} < a_1\sum_{k=2}^{[x]-1}\frac{1}{\ln k} < \frac{a_1}{\ln 2} + a_1\int_2^x\frac{dt}{\ln t}$$

$$= \frac{a_1}{\ln 2} + a_1\left\{\int_2^{\sqrt{x}}\frac{dt}{\ln t} + \int_{\sqrt{x}}^x\frac{dt}{\ln t}\right\}$$

$$< \frac{a_1}{\ln 2} + \frac{a_1}{\ln 2}\sqrt{x} + a_1\frac{x}{\ln x} < a_2\pi(x),$$

where a_1, a_2 are positive constants, and the inequality $\sqrt{x} < \frac{x}{\ln x}$ was used in the last step (why?). Furthermore,

$$\ln[x] > \ln(x-1) = \ln x + \ln\left(1 - \frac{1}{x}\right)$$

$$> \ln x - \frac{1}{x-1}.$$

Equation 5.37 follows from the above three relations.

Next, let us prove Equation 5.38. We know that

$$\Psi(x) = \sum_{n \le x} \Lambda(n) = \sum_{p^a \le x} \ln p,$$

where the right-most sum is taken over prime p and positive integer a such that $p^a \le x$. It is easy to see that when p is fixed, the range of a is $1 \le a \le \dfrac{\ln x}{\ln p}$. By denoting $a_p = \frac{\ln x}{\ln p}$, we have

$$\Psi(x) = \sum_{p \le x} \ln p + \sum_{p^a \le x; a \ge 2} \ln p$$

$$= \theta(x) + \sum_{p \le \sqrt{x}} \ln p \sum_{2 \le a \le a_p} 1$$

$$\le \theta(x) + \sum_{p \le \sqrt{x}} \ln p \cdot \frac{\ln x}{\ln p}$$

$$\le \theta(x) + x^{\frac{1}{2}} \ln x.$$

This yields (5.38).

Theorem 5.17 explains why $\theta(x)$ and $\Psi(x)$ are introduced to replace $\pi(x)$ to study the prime distribution. It is stated as follows.

Theorem 5.17 *Let $x \ge 2$.*

(I) *The following three items are equivalent:*

1. *There are positive constants d_1, d_2 such that*

$$d_1 \frac{x}{\ln x} < \pi(x) < d_2 \frac{x}{\ln x}.$$

2. *There are positive constants d_3, d_4 such that*

$$d_3 x < \theta(x) < d_4 x.$$

3. *There are positive constants d_5, d_6 such that*

$$d_5 x < \Psi(x) < d_6 x.$$

(II) *The following three items are equivalent:*

4. $\lim\limits_{x \to \infty} \dfrac{\pi(x) \ln x}{x} = 1.$

5. $\lim\limits_{x \to \infty} \dfrac{\theta(x)}{x} = 1.$

6. $\lim\limits_{x \to \infty} \dfrac{\Psi(x)}{x} = 1.$

CHAPTER 6

Simple Continued Fractions

R EPRESENTATION OF numbers is an important topic in the study of number theory. Expressing a number in terms of a continued fraction is of particular interest among the different ways of representing numbers. It provides knowledge of the nature and properties of numbers from a new perspective. Combined with other methods, continued fractions have been used in solving many difficult problems arising from the course of understanding numbers. From examples, continued fraction methods play important roles in the rational approximation of real numbers, finding roots of quadratic equations, and solving Diophantine equations and congruence equations. These provide useful ideas and methods for the development of number theory. In this chapter, we introduce basic knowledge of continued fractions and an application in cryptography, that is, a method of attacking the Rivest, Shamir, and Aldeman (RSA) encryption algorithm using continued fractions.

6.1 SIMPLE CONTINUED FRACTIONS AND THEIR BASIC PROPERTIES

We call the expression of the following form

$$a_1 + \cfrac{b_1}{a_2 + \cfrac{b_2}{a_3 + \cfrac{b_3}{a_4 + \ddots}}}$$

a continued fraction. In general, numbers $a_1, a_2, a_3, \cdots, b_1, b_2, b_3, \cdots$ can be either real or complex numbers, and the number of terms can be either finite or infinite. In this chapter, however, we will restrict ourselves to the discussion of simple continued fractions—the continued fractions with the form

$$a_1 + \cfrac{1}{a_2 + \cfrac{1}{a_3 + \cfrac{1}{a_4 + \ddots}}}, \tag{6.1}$$

where a_1 is an integer and $a_2, a_3, a_4, \cdots$ are positive integers. By a finite simple continued fraction, we mean a continued fraction with only finitely many terms. A simple continued fraction is called an infinite simple continued fraction if it is not a finite simple continued fraction. A simple continued fraction is often written in an abbreviated notation $[a_1, a_2, \cdots, a_k, \cdots]$.

Definition 6.1 For $k \geq 1$,

$$[a_1, a_2, \cdots, a_k] = a_1 + \cfrac{1}{a_2 + \cfrac{1}{a_3 + \cfrac{1}{\ddots + \cfrac{1}{a_k}}}}$$

is called the kth convergent of the simple continued fraction $[a_1, a_2, \cdots, a_k, \cdots]$. □

This concept applies to both finite simple continued fractions and infinite simple continued fractions. The difference is that a finite simple continued fraction has only finitely many convergents, but the convergents for an infinite simple continued fraction form an infinite set. Denote

$$[a_1, a_2, \cdots, a_k] = \frac{p_k}{q_k},$$

then we see that $\frac{p_k}{q_k}$ is a function of $a_1, a_2, \cdots, a_k$. It is easy to verify that

$$\frac{p_1}{q_1} = \frac{a_1}{1}, \quad \frac{p_2}{q_2} = \frac{a_2 a_1 + 1}{a_2}, \quad \frac{p_3}{q_3} = \frac{a_3(a_2 a_1 + 1) + a_1}{a_3 a_2 + 1}.$$

In general, we have the following.

Theorem 6.2 *If $\frac{p_1}{q_1}, \frac{p_2}{q_2}, \cdots, \frac{p_k}{q_k}, \cdots$ are convergents of the simple continued fraction $[a_1, a_2, \cdots, a_k, \cdots]$, then the numerators and denominators of these convergents satisfy the following recursive relations:*

$$p_1 = a_1, p_2 = a_2 a_1 + 1, p_k = a_k p_{k-1} + p_{k-2},$$
$$q_1 = 1, q_2 = a_2, q_k = a_k q_{k-1} + q_{k-2},$$

for $k \geq 3$.

Proof We prove this by induction on k. The result is obvious when $k = 1, 2, 3$. If the result holds true for positive integers less than k, then

$$\frac{p_k}{q_k} = [a_1, a_2, \cdots, a_{k-1}, a_k] = \left[a_1, a_2, \cdots, a_{k-1} + \frac{1}{a_k}\right]$$
$$= \frac{(a_{k-1} + \frac{1}{a_k})p_{k-2} + p_{k-3}}{(a_{k-1} + \frac{1}{a_k})q_{k-2} + q_{k-3}} = \frac{a_k(a_{k-1}p_{k-2} + p_{k-3}) + p_{k-2}}{a_k(a_{k-1}q_{k-2} + q_{k-3}) + q_{k-2}}.$$

From the induction hypothesis, $p_{k-1} = a_{k-1}p_{k-2} + p_{k-3}, q_{k-1} = a_{k-1}q_{k-2} + q_{k-3}$. We plug these relations in the above formula and get

$$p_k = a_k p_{k-1} + p_{k-2}, q_k = a_k q_{k-1} + q_{k-2}.$$

The theorem is proved.

Theorem 6.3 *If the kth convergent of the simple continued fraction $[a_1, a_2, \cdots, a_k, \cdots]$ is $\frac{p_k}{q_k}, k = 1, 2, \cdots, n, \cdots$, then the following two relations hold:*

$$p_k q_{k-1} - p_{k-1} q_k = (-1)^k \ (k \geqslant 2),$$
$$p_k q_{k-2} - p_{k-2} q_k = (-1)^{k-1} a_k \ (k \geqslant 3).$$

Proof We prove this by induction on k. When $k = 2$, we have

$$p_2 q_1 - p_1 q_2 = (a_2 a_1 + 1) - a_1 a_2 = 1 = (-1)^2.$$

Assume that $p_{k-1}q_{k-2} - p_{k-2}q_{k-1} = (-1)^{k-1}$, then by Theorem 6.2,

$$p_k q_{k-1} - p_{k-1} q_k = (a_k p_{k-1} + p_{k-2})q_{k-1} - p_{k-1}(a_k q_{k-1} + q_{k-2})$$
$$= p_{k-2}q_{k-1} - p_{k-1}q_{k-2} = -(-1)^{k-1} = (-1)^k.$$

Therefore, the first relation holds.

From the first relation and Theorem 6.2, we see that

$$p_k q_{k-2} - p_{k-2} q_k = (a_k p_{k-1} + p_{k-2}) q_{k-2} - p_{k-2}(a_k q_{k-1} + q_{k-2})$$
$$= a_k(p_{k-1} q_{k-2} - p_{k-2} q_{k-1}) = (-1)^{k-1} a_k.$$

The theorem is proved.

Corollary 6.4 *The rational numbers* $\frac{p_k}{q_k}, k = 1, 2, \cdots$ *are in reduced form, and*

$$\frac{p_k}{q_k} - \frac{p_{k-1}}{q_{k-1}} = \frac{(-1)^k}{q_k q_{k-1}} \ (k \geqslant 2),$$
$$\frac{p_k}{q_k} - \frac{p_{k-2}}{q_{k-2}} = \frac{(-1)^{k-1} a_k}{q_k q_{k-2}} \ (k \geqslant 3).$$

Remark If the numbers a_i in (6.1) are real numbers (or even complex numbers), the formulas in Theorems 6.2 and 6.3 still hold.

By Theorem 6.2 and induction, it is not difficult to get an estimation of the lower bound of q_k.

Theorem 6.5 *If* $k \geq 3$, *then* $q_k \geq 2^{\frac{k-1}{2}}$.

We can see that from the definitions of q_1, q_2, for any $k \geq 1, q_k \geq k - 1$. Combining this fact with Corollary 6.4, it is easy to check the following inequalities:

$$\frac{p_{2(k-1)}}{q_{2(k-1)}} > \frac{p_{2k}}{q_{2k}}, \ \frac{p_{2(k+1)}}{q_{2(k+1)}} > \frac{p_{2(k-1)}}{q_{2(k-1)}}, \ \frac{p_{2k}}{q_{2k}} > \frac{p_{2(k-1)}}{q_{2(k-1)}}.$$

For an infinite simple continued fraction, if the limit of $\frac{p_k}{q_k}$ exists as $k \to \infty$, then this limit is defined to be the value of the continued fraction. The main theorem of this section states the following.

Theorem 6.6 *Every simple continued fraction is a real number.*

Proof It is obvious that every finite simple continued fraction is a rational number. It remains to consider the case of infinite simple continued fractions. Let

$$[a_1, a_2, \cdots, a_k, \cdots]$$

be an arbitrary infinite simple continued fraction with convergents $\frac{p_k}{q_k}$, $k = 1, 2, \cdots$. From the discussion above, we know that

$$\frac{p_1}{q_1}, \frac{p_3}{q_3}, \cdots, \frac{p_{2(k-1)}}{q_{2(k-1)}}, \cdots$$

is a bounded increasing sequence, and

$$\frac{p_2}{q_2}, \frac{p_4}{q_4}, \cdots, \frac{p_{2k}}{q_{2k}}, \cdots$$

is a bounded decreasing sequence. Furthermore, by Corollary 6.4,

$$0 < \frac{p_{2k}}{q_{2k}} - \frac{p_{2(k-1)}}{q_{2(k-1)}} = \frac{1}{q_{2k}q_{2(k-1)}} \leq \frac{1}{(2k-1)(2k-2)} \to 0,$$

so $\left[\frac{p_{2k-1}}{q_{2k-1}}, \frac{p_{2k}}{q_{2k}}\right]$ $(k = 1, 2, \cdots)$ form a set of nested closed intervals; hence, $\lim_{k \to \infty} \frac{p_k}{q_k}$ exists, and the theorem is proved.

Example 6.7 *Find the value of infinite continued fraction* $[1, 1, 1, \cdots, 1, \cdots]$.

Solution: Denote $x = [1, 1, 1, \cdots, 1, \cdots]$. Then

$$x = 1 + \frac{1}{[1, 1, 1, \cdots, 1, \cdots]} = 1 + \frac{1}{x}.$$

This shows that x satisfies the equation

$$x^2 - x - 1 = 0.$$

Since $x > 0$,

$$x = \frac{1 + \sqrt{5}}{2}.$$

6.2   SIMPLE CONTINUED FRACTION REPRESENTATIONS OF REAL NUMBERS

In Section 6.1, we introduced technical notations of continued fractions and proved that every simple continued fraction represents a unique real number. In order to demonstrate the particular advantages of continued fraction representation of numbers, we must solve the problem of whether every real number can be uniquely represented by a simple continued fraction. In this section, it is theoretically proved that every real number basically has a unique simple continued fraction representation. An application of convergent representation of a simple continued fraction in the rational approximation of irrational numbers is also presented.

Theorem 6.8 *Any real number α can be represented as a simple continued fraction. Furthermore, if α is an irrational number, then the representation is unique; if α is a rational number, then its simple continued fraction has finitely many terms and the representation is unique if its last term is greater than 1.*

Proof Let α be a given real number. If α is rational, then we set $\alpha = \dfrac{a}{b}, b > 0$. By the Euclidean division algorithm, we have

$$\frac{a}{b} = q_1 + \frac{r_1}{b}, \quad 0 < \frac{r_1}{b} < 1,$$

$$\frac{b}{r_1} = q_2 + \frac{r_2}{r_1}, \quad 0 < \frac{r_2}{r_1} < 1, \quad q_2 \geqslant 1,$$

$$\cdots\cdots\cdots\cdots$$

$$\frac{r_{n-2}}{r_{n-1}} = q_n + \frac{r_n}{r_{n-1}}, \quad 0 < \frac{r_n}{r_{n-1}} < 1, \quad q_n \geqslant 1,$$

$$\frac{r_{n-1}}{r_n} = q_{n+1}, \quad q_{n+1} > 1.$$

This shows

$$\alpha = \frac{a}{b} = [q_1, q_2, \cdots, q_{n+1}], q_{n+1} > 1,$$

namely, every rational number can be represented as a simple continued fraction. By the uniqueness of the Euclidean algorithm and the definition of simple continued fraction, the integers $q_1, q_2, \cdots, q_{n+1}$ are uniquely determined. Therefore, if the last term of the continued fraction is greater than 1, the expression is unique. But when $q_{n+1} > 1$,

$$\frac{1}{q_{n+1}} = \frac{1}{(q_{n+1} - 1) + \frac{1}{1}},$$

and this implies that

$$\alpha = \frac{a}{b} = [q_1, q_2, \cdots, q_{n+1}] = [q_1, q_2, \cdots, q_{n+1} - 1, 1].$$

So any rational number has at most two simple continued fraction representations.

If α is an irrational number, then from $\alpha = [\alpha] + \{\alpha\}, 0 < \{\alpha\} < 1$ we get

$$\alpha = a_1 + \frac{1}{\alpha_1}, \quad a_1 = [\alpha], \quad \alpha_1 = \frac{1}{\{\alpha\}} > 1,$$

$$\alpha_1 = a_2 + \frac{1}{\alpha_2}, \quad a_2 = [\alpha_1], \quad \alpha_2 = \frac{1}{\{\alpha_1\}} > 1,$$

$$\cdots \cdots \cdots \cdots$$

$$\alpha_{k-1} = a_k + \frac{1}{\alpha_k}, \quad a_k = [\alpha_{k-1}], \quad \alpha_k = \frac{1}{\{\alpha_{k-1}\}} > 1.$$

$$\cdots \cdots \cdots \cdots$$

According to the above procedure, α can be expanded as an infinite simple continued fraction $[a_1; a_2, \cdots, a_k, \cdots]$. By Theorem 6.6, this infinite simple continued fraction must converge to a real number. We prove that this real number is α.

Since

$$\alpha = [a_1, \alpha_1] = [a_1, a_2, \alpha_2] = [a_1, a_2, \cdots, a_k, \alpha_k] = \frac{\alpha_k p_k + p_{k-1}}{\alpha_k q_k + q_{k-1}},$$

$$\alpha_k q_k + q_{k-1} > a_{k+1} q_k + q_{k-1} = q_{k+1},$$

then

$$\left| \alpha - \frac{p_k}{q_k} \right| = \left| \frac{\alpha_k p_k + p_{k-1}}{\alpha_k q_k + q_{k-1}} - \frac{p_k}{q_k} \right|$$

$$= \left| \frac{(-1)^{k-1}}{q_k(\alpha_k q_k + q_{k-1})} \right| < \frac{1}{q_k q_{k+1}} < \frac{1}{k(k-1)}.$$

Hence, $\lim_{k \to \infty} \frac{p_k}{q_k} = \alpha$, and therefore,

$$\alpha = [a_1, a_2, \cdots, a_k, \cdots].$$

From the definition of simple continued fractions, it is straightforward to check that this infinite simple continued fraction is unique. The theorem is proved.

Example 6.9 *Use a simple continued fraction to respresent $\alpha = \sqrt{8}$.*

Solution:

$$a_1 = [\alpha] = 2, \quad \alpha_1 = \frac{1}{\sqrt{8}-2} = \frac{\sqrt{8}+2}{4},$$

$$a_2 = [\alpha_1] = 1, \quad \alpha_2 = \frac{1}{\frac{\sqrt{8}+2}{4}-1} = \sqrt{8}+2,$$

$$a_3 = [\alpha_2] = 4, \quad \alpha_3 = \frac{1}{\sqrt{8}+2-4} = \frac{\sqrt{8}+2}{4},$$

$$\cdots\cdots\cdots\cdots$$

These mean that

$$\alpha = \sqrt{8} = [2, 1, 4, 1, 4, \cdots].$$

We now introduce a result of using continued fractions in the rational approximation of a real number.

Theorem 6.10 *Let α be an arbitrary real number, and $\frac{p_k}{q_k}$ be the kth convergent of α, then for any $0 < q \leqslant q_k$,*

$$\left| \alpha - \frac{p_k}{q_k} \right| \leqslant \left| \alpha - \frac{p}{q} \right|.$$

Thus, among the rational numbers with denominator not greater than q_k, $\frac{p_k}{q_k}$ is the best rational approximation of α.

Proof The theorem holds obviously if $\alpha = \frac{p_k}{q_k}$. So we only need to discuss the case that $\alpha \neq \frac{p_k}{q_k}$. In this case, α has the $(k+1)$st convergent $\frac{p_{k+1}}{q_{k+1}}$. Without loss of generality, we assume $\frac{p_k}{q_k} < \frac{p_{k+1}}{q_{k+1}}$ (the discussion for $\frac{p_{k+1}}{q_{k+1}} < \frac{p_k}{q_k}$ is similar).

By Theorems 6.6 and 6.8 and $\frac{p_k}{q_k} < \frac{p_{k+1}}{q_{k+1}}$, we get

$$\frac{p_k}{q_k} < \alpha \leqslant \frac{p_{k+1}}{q_{k+1}}.$$

If $\frac{p}{q} \leqslant \frac{p_k}{q_k}$, the truth of the result is obvious. If $\frac{p_{k+1}}{q_{k+1}} < \frac{p}{q}$, then

$$\left| \alpha - \frac{p}{q} \right| \geqslant \left| \frac{p_{k+1}}{q_{k+1}} - \frac{p}{q} \right| \geqslant \frac{1}{qq_{k+1}} \geqslant \frac{1}{q_k q_{k+1}}.$$

From the proof of Theorem 6.8, we see that

$$\left| \alpha - \frac{p_k}{q_k} \right| < \frac{1}{q_k q_{k+1}} < \left| \alpha - \frac{p}{q} \right|.$$

This means we only need to prove for $0 < q \leqslant q_k$,

$$\frac{p}{q} \leqslant \frac{p_k}{q_k} \quad \text{or} \quad \frac{p_{k+1}}{q_{k+1}} < \frac{p}{q}.$$

If not, assume $\frac{p_k}{q_k} < \frac{p}{q} \leqslant \frac{p_{k+1}}{q_{k+1}}$. Since $\frac{p_{k+1}}{q_{k+1}}$ is in reduced form, and $q \leqslant q_k < q_{k+1}$, we have $\frac{p_k}{q_k} < \frac{p}{q} < \frac{p_{k+1}}{q_{k+1}}$. Therefore,

$$\frac{p}{q} - \frac{p_k}{q_k} = \frac{pq_k - qp_k}{qq_k} \geqslant \frac{1}{qq_k}, \quad \frac{p_{k+1}}{q_{k+1}} - \frac{p}{q} \geqslant \frac{1}{q_{k+1}q}.$$

As $pq_k - qp_k > 0$ and $p_{k+1}q - q_{k+1}p > 0$, from $q_{k+1} + q_k > q$ we get

$$\frac{p_{k+1}}{q_{k+1}} - \frac{p_k}{q_k} \geqslant \frac{q_{k+1} + q_k}{qq_kq_{k+1}} > \frac{1}{q_kq_{k+1}}.$$

This contradicts Corollary 6.4. The theorem is proved.

Example 6.11 *Find a rational approximation of $\sqrt{13}+1$ with four decimal places of accuracy.*

Solution: A routine calculation shows $\sqrt{13}+1 = [4, 1, 1, 1, 1, 6, 1, 1, 1, 1, 6, \cdots]$, so we can get some approximations:

$$4, 5, \frac{9}{2}, \frac{14}{3}, \frac{23}{5}, \frac{152}{33}, \frac{175}{38}, \frac{327}{71}, \frac{502}{109}, \frac{829}{180}, \cdots.$$

By the following inequality obtained in the proof of Theorem 6.10,

$$\left| \alpha - \frac{p_k}{q_k} \right| < \frac{1}{q_kq_{k+1}},$$

we see that

$$\left| \sqrt{13}+1 - \frac{502}{109} \right| < \frac{1}{109 \times 180} < \frac{1}{10^4},$$

so $\frac{502}{109}$ is what we wanted.

Based on the above theorems, we have the following.

Theorem 6.12 *Let α be an arbitrary real number. Then at least one of the two consecutive convergents of α satisfies*

$$\left| \alpha - \frac{p}{q} \right| < \frac{1}{2q^2}.$$

Proof By Theorem 6.10, we may assume $\frac{p_k}{q_k} < \alpha \le \frac{p_{k+1}}{q_{k+1}}$. Thus,

$$\frac{p_{k+1}}{q_{k+1}} - \frac{p_k}{q_k} = \left(\frac{p_{k+1}}{q_{k+1}} - \alpha\right) + \left(\alpha - \frac{p_k}{q_k}\right).$$

Assume that result is not true, then $\frac{p_{k+1}}{q_{k+1}} - \alpha \ge \frac{1}{2q_{k+1}^2}, \alpha - \frac{p_k}{q_k} \ge \frac{1}{2q_k^2}$; therefore,

$$\frac{1}{q_{k+1}q_k} = \frac{p_{k+1}}{q_{k+1}} - \frac{p_k}{q_k} \ge \frac{1}{2q_{k+1}^2} + \frac{1}{2q_k^2},$$

that is, $(q_{k+1} - q_k)^2 \le 0$. This is impossible, and the theorem is proved.

Conversely, one can prove the following.

Theorem 6.13 *Let α be an arbitrary real number. If the rational number $\frac{p}{q}$ satisfies*

$$\left|\alpha - \frac{p}{q}\right| < \frac{1}{2q^2},$$

then $\frac{p}{q}$ must be a convergent of α.

The proof is left to the reader.

6.3 APPLICATION OF CONTINUED FRACTION IN CRYPTOGRAPHY — ATTACK TO RSA WITH SMALL DECRYPTION EXPONENTS

The main purpose of this section is to introduce an application of continued fraction in the analysis of the RSA public key algorithm. More precisely, we use the basic knowledge and theorems from previous sections to design an attack to the RSA algorithm for the case of small decryption keys. We first introduce the RSA public key algorithm.

6.3.1 Introduction to the RSA Public Key Algorithm

In 1977, Rivest, Shamir, and Aldeman proposed the RSA public key algorithm [7], and RSA comes from the first letters of the surnames of the three inventors. This algorithm is one of the most widely used public key algorithms.

We first describe the key generation process of the RSA algorithm: Randomly generate two large prime numbers p and q (usually

p and q are of the same length in terms of binary representation); $N = pq$ is called the RSA modulus. Compute Euler's function $\varphi(N) := (p-1)(q-1)$ for N. Randomly select an integer e with $1 < e < \varphi(N)$, such that $\gcd(e, \varphi(N)) = 1$. Use the extended Euclidean algorithm to compute d such that $ed \equiv 1 \pmod{\varphi(N)}$ and $1 < d < \varphi(N)$. Publish (N, e) but keep d, p, q secret. In other words, the public key of RSA encryption system is (N, e), and the private key is d.

If a user B wants to send an encryption of the message m to the user A, B uses the public key of A to compute $C = m^e \pmod{N}$, then sends C to the user A. After receiving the ciphertext C, A uses private key d to perform decryption $m = C^d \pmod{N}$.

The correctness of decryption: Since the public key exponent e and the private key d satisfy the RSA equation $ed \equiv 1 \pmod{\varphi)(N)}$, there exists integer k such that $ed = 1 + k\varphi(N)$. Let $m \in \mathbb{Z}_N^*$, by Euler's theorem,

$$C^d \equiv m^{ed} \equiv m^{1+k\varphi(N)} \equiv m \pmod{N}.$$

6.3.2 Attacking RSA by Continued Fractions

To speed up the decryption process for RSA, an effective method is to choose a smaller decryption exponent d. For example, if a smart card is communicating with a supercomputer (a server), due to the constraint of computing power, the smart card tends to use a smaller decryption exponent d, while the server uses the bigger encryption exponent e. Weiner [8] first suggested an attack to RSA with small decryption exponent, and proved that if $d < \frac{1}{3}N^{\frac{1}{4}}$, then the RSA modulus N can be efficiently factorized.

Next, we introduce the analysis method of Weiner. First, we assume that the RSA modulus N has two prime factors p and q satisfying $p < q < 2p$. Then we have $p + q < 3\sqrt{N}$. The value $\varphi(N) = N - p - q + 1$ of the Euler's function of N satisfies $N - 3\sqrt{N} < \varphi(N) < N$. From the key generation process of RSA, we know that $ed = 1 + k\varphi(N)$. Therefore,

$$\left| \frac{e}{N} - \frac{k}{d} \right| = \left| \frac{ed - kN}{Nd} \right| = \frac{k(p+q-1)-1}{Nd}.$$

Since $0 < k < d$, $p + q < 3\sqrt{N}$,

$$\left| \frac{e}{N} - \frac{k}{d} \right| < \frac{3}{\sqrt{N}}.$$

If $d < \frac{1}{3}N^{\frac{1}{4}}$, then

$$\left| \frac{e}{N} - \frac{k}{d} \right| < \frac{1}{2d^2}.$$

By Theorem 6.13, we know that $\frac{k}{d}$ is a convergent of the rational number $\frac{e}{N}$.

Thus, when $d < \frac{1}{3}N^{\frac{1}{4}}$, we can get an efficient method to factorize the RSA modulus N: Use the Euclidean algorithm to compute the convergent $\frac{k_i}{d_i}$ of $\frac{e}{N}$ for each i; compute $T_i = N - \frac{ed_i-1}{k_i} + 1$; and determine where N can be factorized by solving quadratic equation $y^2 - T_i y + N = 0$ for the prime p. If not, compute the next convergent $\frac{k_{i+1}}{d_{i+1}}$, until N is factorized.

EXERCISES

6.1 Find a rational approximation of $\cos 36°$ with five decimal places of accuracy.

6.2 Prove Theorem 6.13.

6.3 Prove that the maximum value of the number of iterations i in the continued fraction attack of RSA is less than or equal to $\frac{1}{2} \log N$.

CHAPTER 7

Basic Concepts

T HE MAIN content of modern algebra (abstract algebra) is the algebraic system—a nonempty set—together with one or more binary operations (also called algebraic operations) defined on it. This book will mainly cover three basic algebraic systems—groups, rings, and fields. In this chapter, we introduce some basic concepts that are closely related to these three basic algebraic systems. The concepts include maps between sets, binary operations, homomorphisms and isomorphisms between sets equipped with binary operations, and equivalence relations.

7.1 MAPS

Before introducing the concept of maps, let us briefly recall the symbolic representation of sets and set operations. Usually, we use capital letters $A, B, C, D, \cdots$ to denote sets, and use lowercase letters $a, b, c, d, \cdots$ to represent elements of a set. If a is an element of the set A, then a is said to belong to A, or A contains a, and is denoted by $a \in A$. If a is not an element of A, then a does not belong to A or A does not contain a, and is denoted by $a \notin A$. We use $\emptyset$ to denote the empty set which is a subset of any set. $B \subset A$ denotes that B is a subset of A, and $B \not\subset A$ denotes that B is not a subset of A. $A \cap B$ denotes the intersection of A and B. $A \cup B$ is the union of A and B. $\bar{A}$ denotes the complement of A with respect to the universal set I, that is, $\bar{A} = I - A$. We use $2^A = \{B | B \subset A\}$ denotes the set whose elements are all subsets of A, called the power set of A.

Definition 7.1 Let $A_1, A_2, \cdots, A_n$ be n sets. The set of all ordered tuples $(a_1, a_2, \cdots, a_n)$ with $a_i \in A_i$ for $i = 1, 2, \cdots, n$ is called the Cartesian product of $A_1, A_2, \cdots, A_n$, denoted as $A_1 \times A_2 \times \cdots \times A_n$. □

Next, we introduce a general definition of maps.

Definition 7.2 Let A, B be two given sets. If there is a rule ϕ such that for any element a of A, we get a unique $b \in B$ which corresponds to a, then ϕ is called a map from A to B, and is denoted as $\phi : A \to B$. A is called the domain of ϕ, and B is called the range of ϕ. We say that b is the image of a under ϕ, and a is a pre-image of b, and write $b = \phi(a)$, or express it as

$$\phi : A \longrightarrow B$$
$$a \longmapsto b.$$

□

Definition 7.3 Given a set A,

$$I_A : A \longrightarrow A$$
$$a \longmapsto a, \quad \forall a \in A$$

is a map from A to A, and is called the identity map on A. □

Definition 7.4 Let ϕ and φ be two maps from A to B. If for any $a \in A$, the equality $\phi(a) = \varphi(a)$ holds, then we say that the maps ϕ and φ are equal. □

From Definition 7.4, we can get a simple method to check that two maps are different, namely, finding an $a \in A$, such that $\phi(a) \neq \varphi(a)$.

Definition 7.5 Let ϕ be a map from set A to set B. If for every element b of B, there is an $a \in A$ such that $b = \phi(a)$, then ϕ is called a surjection from A to B.

If for any $a_1 \neq a_2$, $\phi(a_1) \neq \phi(a_2)$ holds, then ϕ is called an injection from A to B.

If ϕ is both surjection and injection, then it is called a bijection. □

By the definition, it is seen that an equivalent statement of ϕ being injection is

$$\phi(a_1) = \phi(a_2) \Rightarrow a_1 = a_2.$$

It can also be seen that there is no bijection between a finite set and its proper subset.

Definition 7.6 Given three sets A, B, C and maps $\phi : A \to B$, $\varphi : B \to C$, for any $a \in A$, the maps ϕ and φ determine the following map

$$\begin{array}{rcc} \eta : & A & \longrightarrow & C \\ & a & \longmapsto & \varphi(\phi(a)). \end{array}$$

η is called the composition of ϕ and φ and is denoted as $\eta = \varphi \circ \phi$. □

Theorem 7.7 Let $\phi : A \to B$, $\varphi : B \to C$, $\eta : C \to D$, then

1. $\eta \circ (\varphi \circ \phi) = (\eta \circ \varphi) \circ \phi$.

2. $I_B \circ \phi = \phi, \phi \circ I_A = \phi$.

Proof

1. By the definition of two maps being equal, it is easy to see that the compositions $\eta \circ (\varphi \circ \phi)$ and $(\eta \circ \varphi) \circ \phi$ have the same domain and range. For any $a \in A$, by the definition of map composition, we have the identity

$$[\eta \circ (\varphi \circ \phi)](a) = \eta[(\varphi \circ \phi)(a)] = \eta[\varphi(\phi(a))] = (\eta \circ \varphi)(\phi(a))$$
$$= [(\eta \circ \varphi) \circ \phi(a)].$$

   The relation is proved.

2. The domains and ranges of $I_B \circ \phi$ and ϕ are all A and B, respectively. For any $a \in A$, we have

$$(I_B \circ \phi)(a) = I_B(\phi(a)) = \phi(a);$$

   that is, $I_B \circ \phi = \phi$. Similarly, we can prove $\phi \circ I_A = \phi$.

Theorem 7.7 means that the composition of maps satisfies the law of association.

Definition 7.8 Let $\phi : A \to B$ be a map. If there is a map $\varphi : B \to A$ such that $\varphi \circ \phi = I_A$, then ϕ is said to be left invertible, and φ is called the left inverse of ϕ. Similarly, if $\phi \circ \varphi = I_B$, then ϕ is said to be right invertible, and φ is called the right inverse of ϕ. If ϕ is both left invertible and right invertible, it is called an invertible map. □

The following theorem provides sufficient and necessary conditions for determining whether a map is left invertible or right invertible.

Theorem 7.9 *Given a map* $\phi : A \to B$,

1. ϕ *is left invertible if and only if* ϕ *is an injection.*

2. ϕ *is right invertible if and only if* ϕ *is a surjection.*

Proof

1. Necessity: Let ϕ be left invertible. Then there is $\varphi : B \to A$ such that $\varphi \circ \phi = I_A$. Assume $\phi(a_1) = \phi(a_2)$, then

$$a_1 = I_A(a_1) = (\varphi \circ \phi)(a_1) = \varphi(\phi(a_1)) = \varphi(\phi(a_2)) = (\varphi \circ \phi)(a_2)$$
$$= I_A(a_2) = a_2.$$

This shows that ϕ is an injection.

Sufficiency: Let ϕ be an injection from A to B. We fix an $a_1 \in A$ and define φ_1 as follows:

$$\varphi_1(b) = \begin{cases} a, & \text{if there is an } a \in A \text{ satisfies } \phi(a) = b, \\ a_1, & \text{if } b \notin \phi(A). \end{cases}$$

φ_1 is a map as for each $b \in B$, $\varphi_1(b)$ is uniquely determined. For any $a \in A$, we have $(\varphi_1 \circ \phi)(a) = \varphi_1(\phi(a)) = \varphi_1(b) = a$, that is, $\varphi_1 \circ \phi = I_A$.

2. Necessity: Let ϕ be right invertible. Then there is $\eta : B \to A$ such that $\phi \circ \eta = I_B$. For any $b \in B$, since

$$b = I_B(b) = (\phi \circ \eta)(b) = \phi(\eta(b)),$$

we know that there exists $\eta(b) \in A$ such that $\phi(\eta(b)) = b$. This shows that ϕ is a surjection.

Sufficiency: Let ϕ be a surjection. Then for each $b \in B$, there is an $a \in A$ such that $\phi(a) = b$. In general, there can be more than one such element a, but we will fix one of them. Then we get a map $\varphi_2 : B \to A$ given by $b \longmapsto a$. For any $b \in B$,

$$(\phi \circ \varphi_2)(b) = \phi(\varphi_2(b)) = \phi(a) = b,$$

namely, $\phi \circ \varphi_2 = I_B$. So ϕ is right invertible.

Corollary 7.10 $\phi : A \to B$ *is an invertible map if and only if ϕ is a bijection.*

If ϕ is a bijection, then it has a left inverse φ and a right inverse η. The next theorem says that φ and η are equal.

Theorem 7.11 *Let* $\phi : A \to B$ *and* $\varphi \circ \phi = I_A$, $\phi \circ \eta = I_B$, *then* $\varphi = \eta$.

Proof By Theorem 7.7,

$$\varphi = \varphi \circ I_B = \varphi \circ (\phi \circ \eta) = (\varphi \circ \phi) \circ \eta = I_A \circ \eta = \eta.$$

The proof is complete.

Definition 7.12 A map from A to A is usually called a transform on A. A surjection, injection, and bijection from A to A are usually called a surjective transform, injective transform, and bijective transform on A, respectively. □

7.2 ALGEBRAIC OPERATIONS

An algebraic system is a set equipped with algebraic operations. Algebraic operations are important in an algebraic system for studying algebraic structures of the set with respect to these operations. In this section, we will use maps to define the concept of algebraic operations. We will also introduce several laws of operations, for example, the associative law, the commutative law, and the distributive law.

Definition 7.13 Let A, B, C be sets. We call a map from $A \times B$ to C an algebraic operation from $A \times B$ to C, and denote the operation as $\circ$. For any $(a, b) \in A \times B$, the result $\circ(a, b) = c$ is also denoted as $a \circ b = c$. □

Example 7.14 *Let* $A = \mathbb{Z}, B = \mathbb{Z} \setminus \{0\}$, *and* $D = \mathbb{Q}$.

$$\begin{aligned} \circ : \quad & A \times B \to D \\ & (a, b) \longmapsto \tfrac{a}{b} = a \circ b, \end{aligned}$$

is an algebraic operation from $A \times B$ to D.

If an algebraic operation ∘ is from $A \times A$ to A, we say that A is closed with respect to ∘, or ∘ is an algebraic operation on A.

Definition 7.15 Let ∘ be an algebraic operation on A. If for $a, b, c \in A$, $(a \circ b) \circ c = a \circ (b \circ c)$ holds, then we say that the algebraic operation ∘ satisfies the associative law, and denote $a \circ b \circ c = (a \circ b) \circ c = a \circ (b \circ c)$. If the associative law does not hold, the expression $a \circ b \circ c$ has no meaning. □

This theorem explains a useful property of the associative law.

Theorem 7.16 *If an algebraic operation ∘ on A satisfies the associative law, then for any n ($n \geq 2$) elements $a_1, a_2, \cdots, a_n$ of A, all $\pi_i(a_1 \circ a_2 \circ \cdots \circ a_n)$ are equal. In other words, the expression $a_1 \circ a_2 \circ \cdots \circ a_n$ is always meaningful.*

This theorem enables us to use the expression $a_1 \circ a_2 \circ \cdots \circ a_n$ freely if the associative law holds. This reflects the importance of the associative law.

Definition 7.17 Let ∘ be an algebraic operation from $A \times A$ to D. If $a \circ b = b \circ a$ for every $(a, b) \in A \times B$, then we say the algebraic operation ∘ satisfies the commutative law. □

Theorem 7.18 *If an algebraic operation ∘ on a set A satisfies both the associative law and the commutative law, then the order of elements in the expression $a_1 \circ a_2 \circ \cdots \circ a_n$ can be changed arbitrarily.*

The proofs of Theorems 7.16 and 7.18 are simple and left to the readers.

It is known than many algebraic systems satisfy the commutative law, but there are exceptions. Consider, for example, the multiplication of matrices of order n and the multiplication of linear transforms. Commutativity is a very important property in algebra. Finally, we take a look at laws that involve two operations—the first distribution law and the second distribution law.

Definition 7.19 Given sets A, B and the following two algebraic operations ⊗ and ⊕,

1. ⊗ is an algebraic operation from $B \times A$ to A.

2. ⊕ is an algebraic operation on A.

If for any $b \in B$ and $a_1, a_2 \in A$ the following holds,

$$b \otimes (a_1 \oplus a_2) = (b \otimes a_1) \oplus (b \otimes a_2),$$

then we say that the algebraic operations $\otimes$ and $\oplus$ satisfy the first distributive law. □

Theorem 7.20 *If $\otimes$ satisfies the associative law, $\otimes$ and $\oplus$ satisfy the first distributive law, then for any element b of B and any elements $a_1, a_2, \cdots, a_n$ of A, we have*

$$b \otimes (a_1 \oplus a_2 \oplus \cdots \oplus a_n) = (b \otimes a_1) \oplus (b \otimes a_2) \oplus \cdots \oplus (b \otimes a_n).$$

The proof of this theorem, which can be done by using induction, is omitted and left as an exercise.

For the second distributive law, the discussion is similar.

Definition 7.21 Given sets A, B and the following two algebraic operations $\otimes$ and $\oplus$:

1. $\otimes$ is an algebraic operation from $A \times B$ to A.

2. $\oplus$ is an algebraic operation on A.

If for any $b \in B$ and $a_1, a_2 \in A$, the following holds,

$$(a_1 \oplus a_2) \otimes b = (a_1 \otimes b) \oplus (a_2 \otimes b),$$

then we say that the algebraic operations $\otimes$ and $\oplus$ satisfy the second distributive law. □

Similarly, we have the following theorem.

Theorem 7.22 *If $\oplus$ satisfies the associative law, $\otimes$ and $\oplus$ satisfy the second distributive law, then for any element b of B and any elements $a_1, a_2, \cdots, a_n$ of A, we have*

$$(a_1 \oplus a_2 \oplus \cdots \oplus a_n) \otimes b = (a_1 \otimes b) \oplus (a_2 \otimes b) \oplus \cdots \oplus (a \otimes b).$$

7.3 HOMOMORPHISMS AND ISOMORPHISMS BETWEEN SETS WITH OPERATIONS

The algebraic operations on a set A discussed in Section 7.2 are special maps from $A \times A$ to A. In this section, we discuss two kinds of maps that are related to algebraic operations—homomorphisms and isomorphisms.

Definition 7.23 Given two sets A and B, let $\circ$ be an algebraic operation on A, $\cdot$ be an algebraic operation on B, and ϕ be a map from A to B. If for any two elements a_1, a_2 of A, the following holds,

$$\phi(a_1 \circ a_2) = \phi(a_1) \cdot \phi(a_2),$$

then ϕ is called a homomorphism from A to B. If the homomorphism ϕ from A to B is a surjection, then ϕ is said to be a surjective homomorphism and is denoted as $A \sim B$. If ϕ is a bijection, then ϕ is said to be an isomorphism from A to B; in this case, A is said to isomorphic to B and is denoted as $A \cong B$. □

An isomorphism not only reflects the one-to-one correspondence between the two sets but also indicates the same algebraic structures on these two sets. If A and B are isomorphic with respect to operations $\circ$ and $\cdot$, then the algebraic systems A and B have no difference from an abstract point of view. If one of the sets has some property related to the algebraic operation on it, then the other set must have a entirely similar property.

Definition 7.24 Let A be an algebraic system with operation $\circ$. An isomorphism from A to A is called an automorphism of A. □

Example 7.25 *Let $m \in \mathbb{N}$ be an odd number, then the map ϕ from $\mathbb{R}$ to $\mathbb{R}$ defined by $x \longmapsto x^m$ is an automorphism of $\mathbb{R}$ with respect to multiplication.*

Example 7.26 *Let g be a primitive root modulo p. Define an operation on the reduced system of residues modulo p as the multiplication modulo p. Then the map f from $\mathbb{Z}_p^*$ to $\mathbb{Z}_{p-1}$ is given by*

$$f : \mathbb{Z}_p^* \to \mathbb{Z}_{p-1},$$
$$a \longmapsto \gamma_{p,g}(a)$$

is an isomorphism.

7.4 EQUIVALENCE RELATIONS AND PARTITIONS

From our early discussion of number theory, we know that a system of residues modulo n actually induces a partition of integers. Sometimes, it is necessary to partition a general set. A partition of a set is closely related to an equivalence relation. First, let us take a look at the concept of binary relation.

Definition 7.27 A subset R of $A \times B$ is called a binary relation between A and B. If $(a, b) \in R$, then we say that a and b are related and it is denoted as aRb; if $(a, b) \notin R$, we say that a and b are not related and it is denoted as $aR'b$. Any subset R of $A \times A$ is called a binary relation on A. □

Equivalence relation is a special binary relation.

Definition 7.28 If $R \subseteq A \times A$ satisfies the following conditions,

1. Reflexivity: $(a, a) \in R$

2. Symmetry: If $(a, b) \in R$, then $(b, a) \in R$

3. Transitivity: If $(a, b) \in R$ and $(b, c) \in R$, then $(a, c) \in R$

then R is called an equivalence relation. □

If R is an equivalence relation and $(a, b) \in R$, then a is said to equivalent to b, denoted by $a \sim b$. If R is an equivalence relation on A, then the set $\bar{x} = \{y | y \in A, (x, y) \in R\}$ is called an equivalence class determined by x.

Property 1 If R is an equivalence relation on A, then for any $x, y \in A$, we have

$$\text{either } \bar{x} = \bar{y} \quad \text{or} \quad \bar{x} \cap \bar{y} = \emptyset.$$

The proof of this property is left as an exercise.

Definition 7.29 If $\{B_i, i \in I\}$ is a set of subsets of A that satisfies the following two conditions:

1. $A = \bigcup_{i \in I} B_i,$

2. $B_i \bigcap B_j = \emptyset, \ \forall i, j \in I, i \neq j,$

then $\{B_i, i \in I\}$ is called a partition of A. □

We have the following result for the notions of partition and equivalence relation.

Theorem 7.30 *A partition of set A determines an equivalence relation on A; conversely, an equivalence relation $\sim$ on set A determines a partition of A.*

Proof

1. Given a partition $\{B_i, i \in I\}$ of A, we define a relation as follows:

$$a \sim b \Leftrightarrow a, b \in B_i$$

for some subset B_i in the partition of A. It is easy to check that $\sim$ is an equivalence relation.

2. Given an equivalence relation $\sim$ on A, we can get a set of equivalence classes $\{\bar{a}, a \in A\}$. By property 1, $\{\bar{a}, a \in A\}$ is a partition of A. The theorem is proved.

Example 7.31 *A congruence relation is an equivalence relation. That is, given modulus m, the congruence relation "$\equiv$" modulo m satisfies three conditions for an equivalence relation.*

EXERCISES

7.1 If $A = \cup_{i=1}^{\infty} A_i$, then for each $i = 1, 2, \cdots$, there is a subset B_i of A_i such that $A = \cup_{i=1}^{\infty} B_i$, and for any $i \neq j$, $B_i \cap B_j = \emptyset$ holds.

7.2 Define two operations $\circ, \cdot$ on $\mathbb{R}$ by

$$a \circ b = a + 2b,$$

and

$$a \cdot b = a - b.$$

Are they satisfying the associative law and the commutative law?

7.3 Let $A = \{a, b, c\}$. Please construct a binary operation on A.

7.4 Let R_1, R_2 be two equivalence relations on A. Is $R_1 \cap R_2$ a binary relation on A? Is it an equivalence relation? Why? Is $R_1 \cup R_2$ a binary relation on A?

7.5 Let $(k, p-1) = 1$. Prove that

$$
\begin{array}{rccc}
f: & \mathbb{Z}_p^* & \longrightarrow & \mathbb{Z}_p^* \\
& a & \longmapsto & a^k
\end{array}
$$

is an automorphism on $\mathbb{Z}_p^*$.

7.6 Let $A = \mathbb{Q}$ with addition as its algebraic operation. Let $A' = \mathbb{Q} \setminus \{0\}$ with multiplication as its algebraic operation. Prove that there is no isomorphism between A and A'. (Hint: Determine the image of 0 first.)

7.7 Suppose that the relation R satisfies symmetry and transitivity. The following shows R also satisfies reflexivity. The argument is as follows: As R satisfies symmetry, $aRb \Rightarrow bRa$; then from transitivity, $aRb, bRa \Rightarrow aRa$. What is wrong in this proof?

CHAPTER 8

Group Theory

W E JUST discussed some basic concepts of modern algebra in Chapter 7. With these preliminaries, we are able to introduce an algebraic system with one algebraic operation—the group. Group theory is an important topic of study in modern algebra. It has a great application in cryptography, especially in public key cryptography. In this chapter, the concepts and related properties of groups, cyclic groups and subgroups, will be given. We will also introduce an important tool in the study of groups—the fundamental homomorphism theorem.

8.1 DEFINITIONS

Let us first introduce the concept of group.

Definition 8.1 Let G be a nonempty set and ∘ be a binary operation defined on G. G (or more precisely, $(G, \circ)$) is said to be a group if ∘ satisfies the following conditions:

1. For any $a, b, c \in G$,

$$(a \circ b) \circ c = a \circ (b \circ c)$$

holds, that is, ∘ satisfies associative law.

2. There exists an element e in G such that for each element g of G,

$$e \circ g = g \circ e = g$$

holds.

3. For each element g in G, there exists an element g' of G such that

$$g \circ g' = g' \circ g = e$$

holds.

It is easy to see that the element e is unique provided that it satisfies $e \circ g = g \circ e = g$ for each element $g \in G$. This element is called the identity of G. For each $g \in G$, the element g' that has the property $g \circ g' = g' \circ g = e$ is also unique. This element is called the inverse of g and is usually denoted as g^{-1}. □

We usually use the symbol $\circ$ or $\cdot$ to denote the only algebraic operation for a group. For convenience, sometimes it is directly denoted as the ordinary addition or multiplication. We can even simply write ab for the operation of a and b without using any symbol, and call the algebraic operation as multiplication.

After defining groups, let us see some examples.

Example 8.2 *The set of all nonzero rational numbers is a group with respect to the ordinary multiplication. The identity is 1 and the inverse of a is $\frac{1}{a}$.*

Example 8.3 *Let $n \in \mathbb{Z}$, the residues classes $\mathbb{Z}_n$ modulo n forms a group with respect to the addition modulo n.*

Definition 8.4 If the number of elements in a group is a finite integer, then the group is said to be a finite group; otherwise the group is said to be an infinite group. The number of elements in a finite group G is called the order of this group and is denoted by $|G|$. □

From the definition, we know that a group satisfies the associative law. However, a group does not necessarily satisfy the commutative law.

Definition 8.5 Given a group $(G, \circ)$, if for any $a, b \in G, a \circ b = b \circ a$ holds, then this group is called a commutative group (also called an Abelian group). □

Another important concept related to the identity e is defined as follows.

Definition 8.6 Given an element g of group G, the smallest positive integer m such that $g^m = e$ is said to be the order (or period) of g. If no such positive integer exists, then the order of g is said to be infinite. □

The order of g defined here is similar to the exponent $\delta_m(g)$ of g defined in elementary number theory. From the previous discussion, we know that the exponent has the property: for any given integer d, if $g^d \equiv 1 \pmod{m}$, then $\delta_m(g)|d$. We have the similar property for the order of a group element.

Theorem 8.7 *Let m be the order of a, then $a^n = e$ if and only if $m|n$.*

Proof Assume $m|n$, then there is an integer k such that $n = mk$. Therefore,
$$a^n = a^{mk} = (a^m)^k = e^k = e.$$
Conversely, assume $a^n = e$ but $m \nmid n$. Then we have $n = mk + r$, $1 \le r < m$. Thus,
$$e = a^n = a^{mk+r} = ea^r = a^r,$$
and this contradicts to the fact that m is the order of a.

In fact, as mentioned before, the reduced system of residues forms a finite group with respect to the multiplication of residue classes, and the exponent of an element is exactly the order of the same element.

Example 8.8 *Let a, b be two elements of a commutative group G. Assume that the orders of a and b are p and q, respectively, with p, q being distinct primes, then the order of ab is pq.*

The proof is similar to that for the exponents modulo m, and is left to the readers.

8.2 CYCLIC GROUPS

We defined groups in Section 8.1. In this section, we introduce an important class of groups—the cyclic groups—and focus on the study of the structure of cyclic groups. The study of group structures is the main goal of group theory. So far, structures of very few classes of groups are well understood. For most groups, further study of their structures is needed. In this section, we will use multiplication for the algebraic operation.

Definition 8.9 If every element of a group G is a power of a fixed element a, that is, $G = \{a^n | n \in \mathbb{Z}\}$, then G is called a cyclic group. We say that G is generated by the element a and is denoted as $G = (a)$. a is called a generator of G. □

Let us have two examples of cyclic groups.

Example 8.10 $G = (\mathbb{Z}, +)$ *is a cyclic group since* $G = (1)$.

Example 8.11 *Let p be a prime number. Then the reduced system of residues $(\mathbb{Z}_p^*, \times)$ modulo p forms a cyclic group. A primitive root g modulo p is a generator of the group.*

Before discussing the strictures of cyclic groups, we rephrase the concept of group isomorphism.

Definition 8.12 Let G, G' be two groups, if there is a map $f : G \to G'$, such that

$$f(ab) = f(a)f(b)$$

holds for all $a, b \in G$, then f is said to be a homomorphism from G to G'. If f is a surjection, then it is called a surjective homomorphism and is denoted as $G \sim G'$; in this case, G' is called the image of f. If f is an injection, then it is called an injective homomorphism. If f is a bijection, then it is called an isomorphism and is denoted as $G \cong G'$; we also say that these two groups are isomorphic in this case. □

By the following theorem, we know that there are only two classes of cyclic groups; Examples 8.10 and 8.11 are two concrete representatives of these two classes.

Theorem 8.13 *Suppose that G is a cyclic group generated by element a. If the order of a is infinite, then G is isomorphic to the additive group of integers; if the order of a is a positive integer n, then G is isomorphic to the additive group of residue classes modulo n.*

Proof If the order of a is infinity, we define $\phi : G \to \mathbb{Z}$ by

$$\varphi(a^k) = k.$$

First, we show that ϕ is a map from G to $\mathbb{Z}$, that is, $a^h = a^k \Rightarrow h = k$, if $a^h = a^k$ but $h \neq k$. We may assume $h > k$, then we get $a^{h-k} = e$, which contradicts the assumption that a has an infinite order. Therefore, ϕ is a map from G to $\mathbb{Z}$. Since

$$\phi(a^h a^k) = \phi(a^{h+k}) = h + k = \phi(a^h)\phi(a^k),$$

ϕ is a homomorphism. It is easy to verify that ϕ is a bijection; hence, G and the additive group $(\mathbb{Z}, +)$ are isomorphic.

If the order of a is a positive integer n, we define φ by

$$\varphi(a^h) = h. \pmod{n}.$$

We prove that φ is an isomorphism from G to $(\mathbb{Z}_n, +)$. By properties of residue classes,

$$a^h = a^k \Leftrightarrow h \equiv k \pmod{n},$$

so φ defines a map that is also an injection. It is easy to see that φ is a surjection, and hence, φ is a bijection. Since

$$\varphi(a^h a^k) = \varphi(a^{h+k}) = (h+k) \pmod{n} = h \pmod{n} + k \pmod{n},$$

we conclude that φ is an isomorphism from G to $(\mathbb{Z}_n, +)$. The theorem is proved.

Thus, we have a complete understanding the structures of cyclic groups. For general groups with complicated structures, it is hard to get a perfect result compared with that of the cyclic groups.

8.3 SUBGROUPS AND COSETS

We have learned the concept of subsets from set theory. In group theory, subgroup is an important concept.

Definition 8.14 Let H be a nonempty subset of group $(G, \circ)$. If H is a group with respect to the operation $\circ$ of G, then H is said to be a subgroup of G and is denoted by $H \leq G$. □

A given group G has at least two subgroups: G and $\{e\}$. These two subgroups are called trivial subgroups; other subgroups of G are called proper subgroups.

Example 8.15 *Let $\mathbb{C}^*$ be the set of nonzero complex numbers, and $G = \{x | x^m = 1, m \in \mathbb{N}, x \in \mathbb{C}^*\}$. Then for a fixed positive integer n, $H = \{x | x^n = 1, x \in \mathbb{C}^*\}$ is a subgroup of G.*

The definition of subgroups gives a method to determine whether a subset is a subgroup. We will describe a simpler method for the same purpose that does not require checking whether the subset H satisfies all conditions for a group.

Theorem 8.16 *Let H be a nonempty set of group G. Then H is a subgroup of G if and only if $ab^{-1} \in H$ for all $a, b \in H$.*

Proof The necessity is obvious. Let us prove the sufficiency. As H is nonempty, so H contains an element a, thus

$$aa^{-1} = e \in H.$$

From $e, a \in H$ we get $ea^{-1} = a^{-1} \in H$. Similarly, if $a, b \in H$, then $b^{-1} \in H$, hence

$$a(b^{-1})^{-1} = ab \in H.$$

This shows that H is a subgroup.

With the concept of subgroups, we can discuss the structure of subgroups of cyclic groups.

Theorem 8.17 *Subgroups of a cyclic group are still cyclic groups.*

Proof Let a be a generator of the cyclic group. If the subgroup H contains only one element, it is certainly a cyclic group. If $H \neq \{e\}$, then since H is nonempty, there must be $a^k \in H$ with $k > 0$; that is, H contains some powers of a. Let $A = \{k \mid k$ is a positive integer such that $a^k \in H\}$. Then A is not empty and hence has a least element, say r. Now for any $a^l \in H$, write $l = qr + s, 0 \leq s < r$. From $a^l = (a^r)^q a^s$, we see that $a^s \in H$, and hence $s \in A$. This forces $s = 0$. Therefore, for any $a^l \in H$, we must have $r \mid l$, that is, $H = (a^r)$.

Theorem 8.18 *If $G = (a)$ is a cyclic group of order n, then for any $a^i \in G$ with $0 \leq i \leq n-1$, the order of the subgroup $H = (a^i)$ is $\frac{n}{(n,i)}$.*

Example 8.19 *If $G = (a)$ is a cyclic group of order n, then it has $\phi(n)$ generators.*

Example 8.20 *If H_i ($i \in I$ for an infinite or a finite index set I) are subgroups of a group G, then $H = \cap_{i \in I} H_i$ is also a subgroup of G.*

Proof Since for each $i \in I$, $e \in H_i$, so H is not empty. For any $a, b \in H$, then $a, b \in H_i$ for each $i \in I$. Since each H_i is a subgroup, ab^{-1} must be H_i. Therefore, $ab^{-1} \in H$ and H is a subgroup.

Let M be a subset of group G and $H_i, i \in I$, are all subgroups of G that contain M, then we can prove that $H = \cap_{i \in I} H_i$ is the least subgroup that contains M. We call this H the subgroup generated by M, and denote it by (M). If M contains only one element, then $H = (M)$ is a cyclic group.

Next, we will define cosets.

Definition 8.21 Let H be a subgroup of group G. For any $a \in G$, the set $aH = \{ah | h \in H\}$ is called a left coset of H, and $Ha = \{ha | h \in H\}$ is called a right coset of H. □

Example 8.22 *For the additive group of integers $(\mathbb{Z}, +)$ and a subgroup $H = \{nk | k \in \mathbb{Z}\}$, we have the following left coset:*

$$aH = \{a + b | b \in H\} = \{a + nk | k \in \mathbb{Z}\}.$$

It is the residue class $\bar{a}$ modulo n.

Theorem 8.23 *Let H be a subgroup of group G. For any cosets aH, bH of H, we have either $aH = bH$, or $aH \cap bH = \phi$.*

Proof If $x \in aH \cap bH$ for some $x \in G$, then there are $h_1, h_2 \in H$, such that $x = ah_1 = bh_2$. For an arbitrary $ah \in aH$,

$$ah = ah_1 h_1^{-1} h = bh_2 h_1^{-1} h = bh' \in bH,$$

so $aH \subseteq bH$. Similarly, we can have $aH \supseteq bH$. Therefore, $aH = bH$ and the theorem is proved.

Example 8.24 *For the following subgroups of the additive group $(\mathbb{Z}_6, +) = \{\bar{0}, \bar{1}, \bar{2}, \bar{3}, \bar{4}, \bar{5}\}$ of residue classes modulo 6:*

$$H_2 = \{\bar{0}, \bar{2}, \bar{4}\}, \quad H_3 = \{\bar{0}, \bar{3}\}.$$

The left cosets of H_2 are

$$\bar{0}H_2 = \bar{2}H_2 = \bar{4}H_2 = H_2, \ \bar{1}H_2 = \bar{3}H_2 = \bar{5}H_2 = \{\bar{1}, \bar{3}, \bar{5}\}.$$

The left cosets of H_3 are

$$\bar{0}H_3 = \bar{3}H_3 = H_3, \ \bar{1}H_3 = \bar{4}H_3 = \{\bar{1}, \bar{4}\}, \ \bar{2}H_3 = \bar{5}H_3 = \{\bar{2}, \bar{5}\}.$$

Theorem 8.25 *Let H be a subgroup of group G. The numbers of left cosets of H is the same as that of the right cosets.*

Proof Let S_L and S_R be the sets of left cosets and right cosets of H, respectively. Define a map φ from S_R to S_L as

$$\varphi(Ha) = a^{-1}H.$$

We will prove ϕ is a bijection from S_R to S_L.

1. ϕ is well defined, that is, any right coset corresponds to only one left coset. In fact, if $Ha = Hb$, then $ab^{-1} \in H$. Since H is a subset, we have $(ab^{-1})^{-1} = ba^{-1} \in H$. This implies $a^{-1}H = b^{-1}H$.

2. An arbitrary element aH of S_L is the image of the element Ha^{-1} of S_R, so ϕ is a surjection.

3. If $Ha \neq Hb$, then $ab^{-1} \notin H$. Since H is a subgroup, so $(ab^{-1})^{-1} = ba^{-1} \notin H$, which means that $a^{-1}H \neq b^{-1}H$. So ϕ is an injection.

Therefore, we have proved that ϕ is a bijection from S_R to S_L.

By Theorem 8.23, group G is the union of all mutually disjoint left cosets of a subgroup H. This forms a partition of G and produces the following equivalence relation: for any $a, b \in G$,

$$a \sim b \Leftrightarrow a^{-1}b \in H.$$

Definition 8.26 The number of right (left) cosets of a subgroup H of a group G is called the index of H in G and is written as $[G : H]$. □

Theorem 8.27 (*Lagrange*) *Let G be a finite group. If H is a subgroup of G, then*

$$[G : 1]| = [G : H][H : 1],$$

where $[G : 1]$ is the index of the subgroup $\{e\}$ in G, that is, the order of G.

Proof Partition G into $[G : H]$ disjoint left cosets of H. Since each left coset contains $|H|$ elements, the conclusion holds.

Theorem 8.28 *The order N of any element a of a finite group G divides the order of G.*

Proof Since a generates a subgroup of G with order n, therefore by Theorem 8.27, n divides the order of G.

Corollary 8.29 (*Euler's theorem*) *Let a, m be positive integers and $(a, m) = 1$. Then*

$$a^{\varphi(m)} \equiv 1 \pmod{m},$$

where $\varphi(m)$ is Euler's function.

Corollary 8.30 *Let G be a cyclic group of order n and $d|n$. Then there is exactly one subgroup of order d.*

Proof We may assume $G = (a)$ and $n = dr$. It is easy to see that the order of a^r is d, thus, (a^r) is a subgroup of G with order d.

Let H be a subgroup of G with order d. Since a subgroup of a cyclic group is still cyclic, we can assume $H = (a^t)$.

There are q, s with $0 \leq s < r$ such that $td = rqd + sd$. So we have $a^{td} = a^{sd} = e$. Note that $sd < rd = n$, we must have $s = 0$. So $t = rq$ and

$$a^t = a^{rq} \in (a^r).$$

This implies that $H \subseteq (a^r)$. Since both H and (a^r) contain d elements, $H = (a^r)$. We have shown that G contains only one subgroup (a^r) of order d.

8.4   FUNDAMENTAL HOMOMORPHISM THEOREM

We have discussed left and right cosets in Section 8.3. In general, for an arbitrary subgroup of the group G, the left coset aH is not necessarily the same as the right coset Ha. However, for a special subgroup of G, a left coset equals the corresponding right coset. We use a special term for the subgroups of this class—invariant subgroups.

Definition 8.31 Let G be a group and H be a subgroup of G. If for any $a \in G$, $aH = Ha$ holds, then H is said to be an invariant subgroup (or normal subgroup of G) and is denoted by $H \triangleleft G$.   □

Example 8.32 *For any group G, its subgroups G and $\{e\}$ are invariant subgroups.*

Besides using the definition, we have several other ways of determining whether a subgroup H of G is invariant.

Theorem 8.33 *Let H be a subgroup of a group G. Then the following four conditions are equivalent:*

(1) *H is an invariant subgroup of G.*

(2) *For any $a \in G$, $aHa^{-1} = H$ holds.*

(3) *For any $a \in G$, $aHa^{-1} \subseteq H$ holds.*

(4) *For any $a \in G$ and $h \in H$, $aha^{-1} \in H$ holds.*

Proof (1) $\Rightarrow$ (2) Since H is invariant, so $aH = Ha$ holds for any $a \in G$. Therefore,

$$aHa^{-1} = (aH)a^{-1} = (Ha)a^{-1} = H(aa^{-1}) = He = H.$$

(2) $\Rightarrow$ (3) is obvious.
(3) $\Rightarrow$ (4) is obvious.
(4) $\Rightarrow$ (1) For any $h \in H$, since $aha^{-1} \in H$, there must be an $h_1 \in H$ such that $aha^{-1} = h_1$, or $ah = h_1 a$. This implies that $aH \subseteq Ha$. Similarly, we get $Ha \subseteq aH$. So $aH = Ha$ holds for any $a \in G$. Thus, H is an invariant subgroup of G.

Let H be an invariant subgroup of a group G. We use $G/H = \{aH | a \in G\}$ to denote the set of cosets of H and define a binary operation $\circ$ on G/H as

$$aH \circ bH = (ab)H.$$

Now we show that the map $\circ$ is well defined: If $aH \circ bH = a'H \circ b'H$, then there exist $h_1, h_2, h_3, h_4 \in H$ such that

$$ah_1 = a'h_2, \quad h_3 b = h_4 b'.$$

This means that $(a')^{-1}ab(b')^{-1} = h_2(h_1)^{-1}(h_3)^{-1}h_4 \in H$. By the property of invariant subgroups, there is an $h \in H$ such that $ab = a'hb' = a'b'(h')$, and hence, $abH = a'b'H$.

Example 8.34 *For the additive group $(\mathbb{Z}, +)$ of integers, $H = \{nk | k \in \mathbb{Z}\}$ is its invariant subgroup. A coset of H is of the form*

$$aH = \{a + nk | k \in \mathbb{Z}\} = \bar{a}.$$

So all cosets form the additive group of residue classes modulo n.

Example 8.35 *Let H be a subgroup of a group G. Then the product of any two left cosets of H is still a left coset if and only if H is an invariant group.*

Proof The sufficiency is obvious. We need to show the necessity. First, we prove $(aH)(bH) = (ab)H$. By the assumption, $(aH)(bH)$ is a left coset, say cH. Since $ab = (ae)(be) \in (aH)(bH)$, we get $ab \in cH$. This means that $abH = cH$. Now for any $h \in H$,

$$(ah)(a^{-1}h) \in (aH)(a^{-1}H) = (aa^{-1})H = H.$$

So H is an invariant subgroup of G.

We have the following important theorem.

Theorem 8.36 *Let H be an invariant subgroup of a group G. Then $(G/H, \circ)$ is a group.*

Proof Since H is an invariant subgroup of G, it is easy to check that the operation $\circ$ on cosets satisfies the associative law. eH is the identity and $x^{-1}H$ is the inverse of xH. In conclusion, $(G/H, \circ)$ is a group.

Definition 8.37 The group formed by the cosets of an invariant subgroup H of a group G is called a quotient group and is simply denoted by G/H. □

It is well known that homomorphism is an important tool for studying groups. Invariant subgroups, quotient groups, and homomorphisms are well connected.

Theorem 8.38 *There is a homomorphism from a group G to any of its quotient group G/H.*

Proof The map $\phi : G \to G/H$ given by $\phi(a) = aH$, $(a \in G)$ is obviously a surjection.

$$\phi(ab) = abH = (aH)(bH) = \phi(aH)\phi(bH)$$

holds, so ϕ is a surjective homomorphism.

In some sense, the converse of the above theorem holds. To explain this, let us define the kernel of a homomorphism.

Definition 8.39 Let G, G' be groups and ϕ be a surjective homomorphism from G to G'. The set of elements of G whose image under ϕ is the identity e' of G' is called the **kernel** of ϕ and is denoted as $\ker \phi$. □

Theorem 8.40 *Let G, G' be groups. If ϕ is a surjective homomorphism from G to G', then $\ker \phi$ is an invariant subgroup of G.*

Proof Denote $K = \ker \phi$. For any $a, b \in K$, from the fact that ϕ is a homomorphism, we get

$$\phi(ab^{-1}) = \phi(a)\phi(b)^{-1} = e'e'^{-1} = e',$$

which shows that K is a subgroup of G. For any $a \in G, k \in K$,

$$\phi(aka^{-1}) = \phi(a)\phi(k)\phi(a)^{-1} = e',$$

that is, $aka^{-1} \in K$. This means that K is an invariant subgroup of G.

Theorem 8.41 *Let G, G' be groups. If ϕ is a surjective homomorphism from G to G', then*

$$G/\ker \phi \cong G'.$$

Proof Denote $\ker \phi = K$. Construct a correspondence φ from G/K to G' as follows: $\varphi(aK) = \phi(a)$ for $a \in G$. We prove that φ is an isomorphism from G/K to G'.

(1) If $aK = bK$, then $b^{-1}a \in K$. Since ϕ is a homomorphism, we have

$$\phi(b^{-1}a) = \phi(b)^{-1}\phi(a) = e',$$

which means that $\varphi(aK) = \varphi(bK)$. This shows that φ is a map from G/K to G'.

(2) Given any $a' \in G'$, there is at least one element $a \in G$ such that $\phi(a) = a'$. From the definition of φ, this means that $aK \in G/K$ is a preimage of a', and hence, φ a surjection from G/K to G'.

(3) If $aK \neq bK$, then

$$\phi(b^{-1}a) = \phi(b)^{-1}\phi(a) \neq e',$$

which means that $\varphi(aK) \neq \varphi(bK)$. So φ is an injection from G/K to G'.

(4) Note that

$$\varphi(aKbK) = \varphi(abK) = \phi(ab) = \phi(a)\phi(b) = \varphi(aK) \cdot \varphi(bK),$$

so φ is a homomorphism from G/K to G'.

Therefore, we get by the above discussion that $G/\ker\phi \cong G$.

The following important theorem is a consequence of Theorems 8.38 and 8.41.

Theorem 8.42 (*The fundamental homomorphism theorem*) *Let G be a group. Then any quotient group of G is an image of homomorphism of G. Conversely, if G' is an image of homomorphism of G (that is, $G' = f(G)$ for some homomorphism f), then $G' \cong G/\ker f$.*

Finally, we mention a property of surjective homomorphism whose proof is left to the reader.

Theorem 8.43 *Let G, G' be groups and ϕ be a surjective homomorphism from G to G'. Then under ϕ,*

1. *The image H' of a subgroup H of G is a subgroup of G'.*

2. *The image N' of an invariant subgroup N of G that contains $\ker\phi$ is an invariant subgroup of G'.*

3. *The preimage H of a subgroup H' of G' is a subgroup of G that contains $\ker\phi$.*

4. *The preimage N of an invariant subgroup N' of G' is an invariant subgroup of G that contains $\ker\phi$.*

By this theorem, we see that there is a one-to-one correspondence between invariant subgroups of G that contain $\ker\phi$ and invariant subgroups of G'.

Example 8.44 *Let G, G' be groups and f be a surjective homomorphism from G to G'. Let H' be an invariant subgroup of G' and*

$$H = f^{-1}(H') = \{a | a \in G, f(a) \in H'\}.$$

Then H is an invariant subgroup of G and $G/H \cong G'/H'$.

Proof By the fundamental homomorphism theorem, there is a surjective homomorphism $\phi : G' \to G'/H'$. Therefore, $\phi \circ f : G \to G'/H'$ is a surjective homomorphism. Denote $\varphi = \phi \circ f$. By the fundamental homomorphism theorem, it suffices to show $\ker \varphi = H$. For any $a \in G$, we have

$$\varphi(a) = (\phi \circ f)(a) = \phi(f(a)) = f(a)H'.$$

If $a \in f^{-1}(H')$, then we have $f(a)H' = H'$ as $f(a) \in H'$, that is, $\varphi(a) = H'$. This means that $a \in \ker \varphi$, or $f^{-1}(H') \subseteq \ker \varphi$.

Conversely, if $a \in \ker \varphi$, then we have $f(a) \in H'$ as $\varphi(a) = f(a)H' = H'$. This means that $a \in f^{-1}(H')$, or $f^{-1}(H') \supseteq \ker \varphi$. Since $\ker \varphi = H$ is an invariant subgroup of G, it is proved that $G/H \cong G'/H'$ by the fundamental homomorphism theorem.

8.5   CONCRETE EXAMPLES OF FINITE GROUPS

We will discuss two concrete examples of finite commutative groups that have significant cryptographic applications. First, let us take a look of the finite group $\mathbb{Z}_n^*$.

Theorem 8.45 *Let $\mathbb{Z}_n^*$ denote the set of reduced residue classes modulo n. For any $\bar{a}, \bar{b} \in \mathbb{Z}_n^*$, the multiplication is defined to be*

$$\bar{a} \times \bar{b} = \overline{a \times b}.$$

Then $(\mathbb{Z}_n^*, \times)$ forms a commutative group of order $\phi(n)$.

Proof If $\bar{a} = \bar{a}', \bar{b} = \bar{b}'$, then $n | a - a'$, $n | b - b'$. So

$$n | (a - a') \times b + a' \times (b - b') = a \times b - a' \times b',$$

or $\overline{a \times b} = \overline{a' \times b'}$. This means that $\times$ is a binary operation. Obviously, $\mathbb{Z}_n^*$ is closed under "$\times$". The associative law holds, and $\bar{1}$ is the identity element. For each $\bar{a} \in \mathbb{Z}_n^*$, its inverse is $\overline{a^{-1}}$, where

$$aa^{-1} \equiv 1 \pmod{n}.$$

For any $\bar{a}, \bar{b} \in \mathbb{Z}_n^*$,

$$\bar{a} \times \bar{b} = \overline{a \times b} = \overline{b \times a} = \bar{b} \times \bar{a}.$$

Therefore, $(\mathbb{Z}_n^*, \times)$ is a finite commutative group as $(\mathbb{Z}_n^*, \times)$ has $\phi(n)$ elements.

Definition 8.46 Let K be a field of numbers. An elliptic curve E is the set of solutions $(x, y) \in K^2$ of the standard cubic curve

$$y^2 + a_1 xy + a_3 y = x^3 + a_2 x^2 + a_4 x + a_6,$$

where coefficients a_i are from K, together with a point at infinity $\mathcal{O}$. $\square$

For some field K, an elliptic curve can be transformed to the following form

$$y^2 = x^3 + ax + b. \tag{8.1}$$

Definition 8.47 The addition operation "+" defined on an elliptic curve (8.1) is as follows: Let $P(x_1, y_1), Q = (x_2, y_2) \in E$ and $\mathcal{O}$ be the point at infinity, then

(1) $P + \mathcal{O} = P$.

(2) If $x_1 = x_2, y_1 = -y_2$, then $P + Q = \mathcal{O}$.

(3) For all other cases, $P + Q = (x_3, y_3)$, where

$$x_3 = \lambda^2 - x_1 - x_2, \quad y_3 = \lambda(x_1 - x_3) - y_1,$$

$$\lambda = \begin{cases} \dfrac{y_2 - y_1}{x_2 - x_1}, & \text{if } P \neq Q, \\[2mm] \dfrac{3x_1^2 + a}{2y_1}, & \text{if } P = Q. \end{cases}$$
$\square$

We will denote $P + P \cdots + P$ (n terms) as nP and set $0P = \mathcal{O}$.

The addition operation described above is of special geometric meaning.

(1) If $x_1 = x_2, y_1 = -y_2$, then the straight line passes P and Q is perpendicular to the x-axis and meets the elliptic curve at the point at infinity $\mathcal{O}$; hence, $P + Q = \mathcal{O}$.

(2) If $x_1 \neq x_2$, then the straight line passes P and Q meets the elliptic curve at the third point $R = (x_3, y_3)$; hence, $P + Q + R = \mathcal{O}$.

(3) For the calculation of $2P$, one needs to find the tangent line at P. This line will intersect with the elliptic curve at another point R that satisfies $P + P + R = \mathcal{O}$. $2P$ is then the symmetry of R with respect to the x-axis.

Theorem 8.48 *The set G of rational points of an elliptic curve is a commutative group with respect to the addition operation.*

The proof is beyond the scope of this book and is therefore omitted.

Example 8.49 *Let E be the elliptic curve (defined on the field of rational numbers) by the equation*

$$y^2 + y = x^3 - x^2$$

and $P = (1, -1) \in E$. Prove that $\{P, 2P, 3P, 4P, 5P = \mathcal{O}\}$ is a group of rational points of E.

Solution: Write the equation as

$$\left(y + \frac{1}{2}\right)^2 = \left(x - \frac{1}{3}\right)^3 - \frac{x}{3} + \frac{1}{27} + \frac{1}{4}.$$

Let $y' = y + 1/2, x' = x - 1/3$, then the equation is simplified to

$$y'^2 = x'^3 - \frac{x'}{3} + \frac{19}{108}.$$

The point $P = (x_1, y_1) = (1, -1)$ becomes $P' = (2/3, -1/2)$ under the above transformation. So

$$2P' = 2\left(\frac{2}{3}, -\frac{1}{2}\right) = \left(-\frac{1}{3}, -\frac{1}{2}\right),$$

and

$$3P' = 3\left(\frac{2}{3}, -\frac{1}{2}\right) = \left(\frac{2}{3}, -\frac{1}{2}\right) + \left(-\frac{1}{3}, -\frac{1}{2}\right) = \left(-\frac{1}{3}, \frac{1}{2}\right),$$

$$5P' = 3\left(\frac{2}{3}, -\frac{1}{2}\right) + 2\left(\frac{2}{3}, -\frac{1}{2}\right) = \left(-\frac{1}{3}, \frac{1}{2}\right) + \left(-\frac{1}{3}, -\frac{1}{2}\right) = \mathcal{O}.$$

Now we claim that $4P' \neq \mathcal{O}$. In fact, from

$$4P' = 2(2P') = \left(\frac{2}{3}, \frac{1}{2}\right),$$

we know that $4P' \neq \mathcal{O}$. This shows that $\{P, 2P, 3P, 4P, 5P = \mathcal{O}\}$ is a group of rational points of E.

EXERCISES

8.1 Let G be the set of invertible matrices of order n over the field of rational numbers. Then G is a group with respect to the operation of matrix multiplication.

8.2 Let G be a group and $u \in G$ a fixed element. Define an operation "$\circ$" on G as
$$a \circ b = au^{-1}b.$$
Prove that $(G, \circ)$ is a group.

8.3 Let n be a fixed natural number and U_n be the set of nth roots of unity, that is, $U_n = \{e^{2k\pi i/n}, k = 0, 1, \cdots, n-1\}$. Then U_n is a cyclic group with respect to the multiplication of complex numbers.

8.4 Suppose that $H \triangleleft G$ and $[G : H] = m$, then for any $x \in G$, $x^m \in H$.

8.5 Let S be a subset of a group G and
$$C(S) = \{a | a \in G, \forall x \in S : ax = xa\}.$$
Then $C(S)$ is a subgroup of G.

8.6 Let G be a cyclic group with generator a, that is, $G = (a)$. Prove that

1. If the order of a is infinite, then $G \cong \mathbb{Z}$.

2. If the order of a is finite, then $G \cong U_n$.

8.7 Suppose that $H \leq K \leq G$, prove
$$[G : H] = [G : K][K : H].$$

8.8 Let A, B be two finite subgroups of a group G. Then

$$|AB| = \frac{|A||B|}{|A \cap B|}.$$

8.9 Let G be a group. Prove that $f : G \to G$ given by $f(x) = x^{-1}$ is an automorphism if and only if G is a commutative group.

8.10 Let A, B be invariant subgroups of a group G. Then $A \cap B$ and AB are both invariant subgroups of G.

8.11 Let U denote the group of all roots of unity. Prove that $\mathbb{Q}/\mathbb{Z}$ and U are isomorphic.

8.12 Let G be a group that has only finitely many subgroups and f be a surjective homomorphism from G to itself. Prove that f must be an automorphism of G.

8.13 The elliptic curve E (over the field of rational numbers) is defined by
$$y^2 + y - xy = x^3.$$

Choose $P = (1, 1) \in E$. Prove that $\{P, 2P, 3P, 4P, 5P, 6P = \mathcal{O}\}$ forms a rational group on E.

CHAPTER 9

Rings and Fields

W E HAVE discussed some fundamental properties of groups in Chapter 8. Now we turn to the discussion of two types of algebraic systems that have two operations—rings and fields. The concepts of rings and fields are not entirely new to us. We have seen some concrete examples of rings and fields in the course of higher algebras, for example, the ring of integers, the field of real numbers, and the field of complex numbers. These also indicate the importance of rings and fields. In this chapter, we discuss the basic concepts of abstract rings and fields and their basic properties in general. We also analyze several important rings and fields. A ring is an algebraic system with two operations that is more complex than groups. However, in the study of some related ring theoretic problems, there are many approaches that are similar to that for groups.

9.1 DEFINITION OF A RING

In many well-known examples of groups, the algebraic operation is usually called multiplication. In fact, the name of an operation is not important; what matters are the structures and properties of the group or other algebraic system with respect to this operation. In a ring, there are two operations with different structures involved, and we need to name them differently to make a distinction. As we will see later, one of the ring operations is quite similar to the addition operation for integers, so we call this operation an addition. The other ring operation will be called multiplication.

Definition 9.1 We call a commutative group an additive group. The operation on this group is called an addition and denoted as +. Under the definition of addition, some corresponding changes on expressions and computing rules are necessary.

1. Since an additive group obeys the associative law, the sum of n elements $a_1, a_2, \cdots, a_n$ makes sense. We use symbol $\sum_{i=1}^{n} a_i$ to denote this sum, that is,

$$\sum_{i=1}^{n} a_i = a_1 + a_2 + \cdots + a_n.$$

   When n is a positive integer, na denotes the sum of adding a n times.

2. The identity of an additive group is called zero and denoted as 0. Obviously, for any element a in an additive group,

$$0 + a = a + 0 = a.$$

3. The unique inverse of an element a is denoted by $-a$ and is called the negative of a. Obviously, $-(-a) = a$. We will also write $a - b$ for $a + (-b)$ and call it a minus b. □

Given the definitions of negative element and "minus," na is well defined whenever n is an integer. In particular, if n is a negative integer, na is the inverse of $|n|a$. We can define multiplication of an integer and an element in the additive group, and have the following well-formed rules:

$$n \cdot a = na, \ (-n)a = -(na),$$

if $n = 0$, $0a = 0$.

Remark In the expression $0a = 0$, the 0 on the left-hand side is the integer zero, while the 0 on the right-hand side is the identity of the additive group.

Under the new notion, for a nonempty set S of an additive group to be a subgroup, a necessary and sufficient condition is

$$\forall a, b \in S \Rightarrow a - b \in S.$$

With the definition of additive group and the related symbols, we are able to define a ring.

Definition 9.2 Let R be a nonempty set with two operations defined on it: the addition $(+)$ and the multiplication $(*)$. If these operations satisfy

1. $(R, +)$ is an additive group.

2. The multiplication obeys the associative law, that is, for any $a, b, c \in R$,
$$a * (b * c) = (a * b) * c.$$

3. Left and right distributive laws of multiplication over addition, that is, for any $a, b, c \in R$,
$$a * (b + c) = a * b + a * c, \quad (b + c) * a = b * a + c * a,$$

then $(R, +, *)$ is said to be a ring, and it can be simply denoted as R. □

In the rest of our discussion, the multiplication symbol is usually omitted—we write ab for $a * b$.

Before getting to other algebraic properties of rings, let us first get familiar with some rules for the two operations. In the ring of integers, we are very familiar with these rules.

1. If $a + b = a + c$, then $b = c$.

2. For any $a \in R$, $0a = a0 = 0$ holds.

3. For any $a, b \in R$, $(-a)b = -ab = a(-b)$.

These properties can be derived immediately from the definition of a ring.

Next, we define the nth power of an element a. In a ring R, a^n stands for
$$a^n = \overbrace{a \cdots a}^{n \text{ terms}}.$$

It is obvious that for positive integers n, m, we have
$$a^n \cdot a^m = a^{m+n}, \quad (a^n)^m = a^{mn}.$$

Finally, we give some examples of rings in order to get more familiar with the definition.

Example 9.3 *The set $\mathbb{Z}$ of all integers is a ring with respect to the addition and the multiplication. It is called the ring of integers.*

Example 9.4 *The system of residues modulo n is a ring with respect to the addition modulo n and multiplication modulo n.*

9.2 INTEGRAL DOMAINS, FIELDS, AND DIVISION RINGS

Given the definition of groups, discussions of various classes of groups that satisfy special conditions (like commutative groups and cyclic groups) are of great importance. For rings, we also need to work with various rings with special properties, just as in group theory. Generally, there are many types of rings. We focus only on rings that have some well-known properties. These properties are usually associated with the multiplication operation; examples include the commutative law, the cancellation law, the existence of inverse, and the existence of identities. Different types of special rings can be defined based on different properties, such as integral domains, division rings, and fields.

Before getting into these special rings, let us introduce some concepts related to the multiplication operation.

The first law of operation in our consideration is the commutative law. This law is not required in the general definition of a ring, and ab is not necessarily ba in a ring. However, commutative law is obeyed by some rings, for example, the ring of integers.

Definition 9.5 A ring R is said to be a commutative ring if for any $a, b \in R$, $ab = ba$ holds. □

It is easy to see that in a commutative ring, for any positive integer n and any elements a, b, we have

$$a^n \cdot b^n = (ab)^n.$$

Definition 9.6 An element e of a ring R is called an identity if for any $a \in R$, $ea = ae = a$ holds. □

In general, a ring may not have an identity. But in this book, we only discuss rings with an identity. Identity plays an important role in the rings. We remark that if a ring R contains an identity, then the identity is unique. In fact, suppose that R has two identities e and e', then

$$e = ee' = e'.$$

In a ring with an identity, we agree that for any $a \in R$, $a \neq 0$, $a^0 = e$.

Example 9.7 *If R has only one element a, then we can define operations as follows:*

$$a + a = a, \ aa = a.$$

In this case, R is a ring.

Usually, we consider a ring R with at least two elements, so R contains at least an element a which is not zero. Since $0a = 0 \neq a$, zero cannot be the identity of R. For rings with identity, we can define a multiplicative inverse of an element.

Definition 9.8 In a ring with identity, a nonzero element b is called an inverse of an element a, if

$$ab = ba = e. \qquad \square$$

It is easy to know that if $a \in R$ has an inverse, then the inverse is unique. We use a^{-1} to denote the inverse of a. The ring of integers has the identity; however, all integers but ± 1 have no inverse.

The cancellation law is an important property for a set with an operation. In rings, the cancellation law and the existence of zero divisors are closely related. We define a zero divisor before describing the cancellation law.

Definition 9.9 Let R be a ring and $a \in R$, $a \neq 0$. If there is an element $b \in R$, $b \neq 0$ such that $ab = 0$, then a is said to be a left zero divisor of R. The definition of a right zero divisor is similar. $\qquad \square$

In a commutative ring, a left zero divisor must be a right zero divisor. However, in a noncommutative ring, a left zero divisor may not be a right zero divisor. There are of course rings that have no zero divisors (left zero divisors and right zero divisors), for example, the ring of integers. It is obvious that in a ring without zero divisors, $ab = 0$ implies that $a = 0$ or $b = 0$.

Example 9.10 *All $n \times n$ matrices over a field F of numbers form a ring with identity with respect to the addition and multiplication of matrices. If $n \geq 2$, this ring is noncommutative and has zero divisors.*

Whether there are divisors in a ring is closely related to the fact of whether it satisfies cancellation law.

Theorem 9.11 *Let R be a ring without zero divisor and $a \in R$, $a \neq 0$. Then $ab = ac$ implies $b = c$. Similarly, $ba = ca$ implies $b = c$.*

Proof If R has no zero divisor, from $ab = ac$ we get $a(b - c) = 0$. Since $a \neq 0$, it forces that $b - c = 0$, that is, $b = c$. Similarly, we can prove $a \neq 0$ and $ba = ca$ imply $b = c$. This shows that two cancellation laws hold in R.

Corollary 9.12 *In a ring, if one cancellation law holds, so does another.*

In the ring of residues classes modulo prime p, we have $p\bar{a} = \bar{0}$ for any $\bar{a}$. In general, for a ring R that has no zero divisor, every nonzero element has the same order in the additive group $(R, +)$. This leads to the following definition.

Definition 9.13 *If R is a ring without zero divisor, the order of the nonzero elements is called the characteristic of R.* □

Theorem 9.14 *If R is a ring without zero divisor and its characteristic is a finite integer n, then n must be a prime.*

Proof If n is not a prime, then there are $n_1 < n$, $n_2 < n$ such that $n = n_1 n_2$. For a nonzero element a of R, $n_1 a \neq 0$, $n_2 a \neq 0$; however,

$$(n_1 a)(n_2 a) = n_1 n_2 a^2 = na^2 = 0.$$

This is contradictory to the assumption that R does not have zero divisors, so the result holds.

We have introduced four possible properties for a ring: Satisfying the cancellation law for multiplication, the existence of identity, the existence of inverse, and the existence of zero divisors. Next, we define special rings that possess part or all of these properties.

Definition 9.15 A ring R is called an integral domain if the following hold:

1. The commutative law holds for multiplication: for any $a, b \in R$, $ab = ba$.

2. R has an identity element e: for any $a \in R$, $ea = ae = a$.

3. R has no zero divisor: for any $a, b \in R$, if $ab = 0$, then $a = 0$ or $b = 0$. □

Putting this simply, an integral domain is a commutative ring with identity and without zero divisors. Integral domains satisfy three of the four properties mentioned earlier. However, the property that every nonzero element has an inverse does not necessarily hold in an integral domain. For example, the ring of integers is an integral domain; however, all elements other than ± 1 have no inverse. Now we present the concept of a field—a ring that possesses all four properties.

Definition 9.16 A ring R with at least two elements is called a field, if

1. R is a commutative ring.

2. R has an identity element.

3. Every nonzero element of R has an inverse. □

It is customary to use letter F for a field. Since every nonzero element of F has an inverse, this means that F has no zero divisor, and hence, fields have all four properties. If we use F^* to denote all nonzero elements of the field F, then by the definition of a field, $(F^*, \cdot)$ forms a group.

There are many examples of fields. The sets of all rational numbers, all real numbers, and all complex numbers are fields with respect to the usual addition and multiplication. These are the field of rational numbers, the field of real numbers, and the field of complex numbers with which we are familiar.

Example 9.17 *If F is a field with four elements, then*

1. *The characteristic of F is 2.*

2. *The elements of F other than 0 or 1 satisfy the equation $x^2 = x - 1$.*

Proof As an additive group F is finite; hence, its characteristic is a prime p and $p|4$. Therefore, the characteristic of F is 2. Let the elements of F other than 0 or 1 be x_1, x_2. Since $(F^*, *)$ is a commutative group, so $x_1 x_2 \in F^*$. We also get $x_1 x_2 = 1$ because none of x_1, x_2 is 1. On the other hand, $(F, +)$ is an additive group of characteristic 2, so $x_1 + x_2 = 1$ (why is the result not 0?). Thus, x_1, x_2 are two roots of the equation $x^2 - x + 1 = 0$.

There is yet another special kind of ring that we call division ring (or skew field). These rings possess all of the properties except for the commutative law.

Definition 9.18 A ring R with at least two elements is called a division ring, if

1. R has an identity element.

2. Every nonzero element of R has an inverse. □

In a division ring, $a^{-1}b$ may not be ba^{-1}, even though they are the same element in a field.

Example 9.19 *Let R be a finite integral domain with the identity 1. Then R is a field.*

Proof It suffices to prove that for any $a \in R \setminus \{0\}$, its inverse a^{-1} exists. We consider the map f from R to R defined by $f(x) = ax$, where x is an arbitrary element of R. Since the cancellation law holds in R, we have $x_1 \neq x_2 \Rightarrow ax_1 \neq ax_2$. If R has n elements, then $f(R) = \{ax | x \in R\}$ also has n elements. This means that $f(R) = R$ and f is a bijection. Therefore, there is an $x \in R$ such that $ax = 1$, that is, $x = a^{-1}$.

Example 9.20 *If p is a prime number, then the ring of residue classes $\mathbb{Z}_p$ modulo p is a field.*

Example 9.21 *Let $F = \{a + b\sqrt{3} | a, b$ are rational numbers$\}$. Then F is a field with respect to the usual addition and multiplication.*

Proof

1. For any $a_1 + b_1\sqrt{3}, a_2 + b_2\sqrt{3} \in F$,

$$(a_1 + b_1\sqrt{3}) + (a_2 + b_2\sqrt{3}) = (a_1 + a_2) + (b_1 + b_2)\sqrt{3} \in F,$$

$$(a_1 + b_1\sqrt{3})(a_2 + b_2\sqrt{3}) = (a_1 a_2 + 3b_1 b_2) + (a_1 b_2 + a_2 b_1)\sqrt{3} \in F,$$

so F is closed with respect to addition and multiplication.

2. It is easy to see that F satisfies the associative laws for multiplication and addition, and the distribution law.

3. The identity element for addition is 0, and the identity element for multiplication is 1.

4. For any $a + b\sqrt{3} \in F$, its negative is $-(a + b\sqrt{3})$, and its multiplicative inverse is $(a - b\sqrt{3})/(a^2 - 3b^2)$ (if a, b are both nonzero). So F is an additive group, and F^* is a multiplicative group. Both $(F, +)$ and $(F^*, \cdot)$ are commutative groups, and hence, F is a field.

9.3   SUBRINGS, IDEALS, AND RING HOMOMORPHISMS

Several special types of ring were discussed in Section 9.2. In this section, we discuss sufficient and necessary conditions for a subset S of a ring R to be a ring under the addition and multiplication of R. Given a subring S of R, S is an additive subgroup of R. Under some suitable conditions, the cosets of S form a quotient ring with respect to addition and multiplication (of cosets). Subrings with such conditions are called ideals. Furthermore, like what we have done for groups, we will discuss the relationship between ring homomorphisms, ideals, and quotient rings.

First, let us define subrings.

Definition 9.22 Let R be a ring and S a subset of R. S is called a subring of R if S itself is a ring with respect to the algebraic operations of R. □

A subset S of a division ring R is called a subdivision ring if S itself is a division ring with respect to the algebraic operations of R. We can also define the concepts of subintegral domain and subfield.

Theorem 9.23 *Let R be a ring and S a nonempty subset of R. Then S is a subring of R if and only if for any $a, b \in S$, $a - b \in S, ab \in S$ hold.*

The proof is left as an exercise.

R itself is a subring of a ring R, the subset $\{0\}$ of R is also a subring. So any nontrivial ring has at least two subrings.

Example 9.24 *All subrings of the ring of residue classes $\mathbb{Z}_6 = \{\bar{0}, \bar{1}, \bar{2}, \bar{3}, \bar{4}, \bar{5}\}$ modulo 6 are $\{0\}$; $\{\bar{0}, \bar{2}, \bar{4}\}$; $\{\bar{0}, \bar{3}\}$; $\mathbb{Z}_6$.*

As an exercise, the readers can prove the intersection $S_1 \cap S_2$ of any two subrings S_1, S_2 of a ring R is a subring. More generally, if $\{S_\alpha\}_{\alpha \in B}$ is a family of subrings of the ring R, where B denotes a family of indexes, then $\cap_{\alpha \in B} S_\alpha$ is a subring of R.

For any nonempty subset T of a ring R, there is always a subring that contains T, for example, R is such a subring. Let $\{S_\alpha | \alpha \in B\}$ be the family of subrings that contain T, then $\cap_{\alpha \in B} S_\alpha$ is the smallest subring of R that contains T. This subring is said to be generated by T and is usually denoted as (T).

Next, we determine the elements of (T). For any $t_1, t_2, \cdots, t_n \in T$, $\pm t_1 t_2 \cdots t_n \in (T)$; thus,

$$\sum \pm t_1 t_2 \cdots t_n \in (T).$$

It is easy to see that $\{\sum \pm t_1 t_2 \cdots t_n | t_i \in T\}$ is a subring of R that contains T and is contained in any subring which contains T. So

$$(T) = \left\{ \sum \pm t_1 t_2 \cdots t_n | t_i \in T \right\}.$$

In particular, if we take $T = \{a\}$, then

$$(T) = \left\{ \sum_{i=1}^{m} n_i a^i | n_i \in \mathbb{Z} \right\}.$$

Let F be a field and S a nonempty set of F. Then the intersection of all subfields of F that contain S is a subfield. This is the smallest subfield that contains S and is called the subfield generated by S. If S is a subring of the field F, then the subfield of F that is generated by S consists of elements of the form ab^{-1}, where $a, b \in S, b \neq 0$.

Example 9.25 *In a ring R, the set of elements that commute with all elements of the ring is a subring I. This subring is called the center of R.*

Proof It suffices to show that for any $x, y \in I$, we have $x - y \in I$, $xy \in I$. In fact, for any $a \in R$,

$$(x - y)a = xa - ya = ax - ay = a(x - y);$$
$$(xy)a = x(ay) = (xa)y = a(xy).$$

So I is a ring.

Obviously, the center of a ring R is a commutative ring. When R is a division ring, its center is a division ring and hence a field.

Let S be a subring of a ring R. The additive subgroup $(S, +)$ of the group $(R, +)$ is invariant. Therefore, the cosets of $(S, +)$ form a quotient group under the cosets addition operation. Now we consider the question about how to define another operation, the multiplication operation, on the quotient group to turn it into a ring. First, we describe a related concept—ideals. Ideals are a very important class of subrings which are in a similar position as the class of invariant subgroups in groups.

Definition 9.26 A nonempty subset A of a ring R is called an ideal subring, or simply an ideal, if

1. For any $a, b \in A$, $a - b \in A$.

2. For any $a \in A, r \in R$, $ra, ar \in A$. □

By the definition, an ideal A is an additive group, and A is closed with respect to the multiplication, so an ideal must be a subring. A nontrivial ring R has at least two ideals—the set contains only zero and R itself. Except for these two ideals, all other ideals of R are called proper ideals.

Example 9.27 *A division ring does not have a proper ideal.*

Proof If A is a nonzero ideal of R, then there is an $a \in A, a \neq 0$. By the definition of an ideal, $a^{-1}a = 1 \in A$. This implies that for any $b \in R$, $b = b \cdot 1 \in A$. So $A = R$.

Therefore, for division rings and fields, the concept of ideals has little use.

Definition 9.28 Let R be a ring and $a \in R$. The smallest ideal of R that contains a is called the ideal generated by a and denoted by (a). We call an ideal generated by a single element is a principal ideal. □

It is not hard to prove that the principal ideal (a) consists of elements of the following form:

$$\sum x_i a y_i + sa + at + na,$$

where x_i, y_i, s, t are arbitrary elements of R and $n \in \mathbb{Z}$.

In some special rings, the elements in a principal ideal can have a simpler form. For example, if R is a commutative ring, then each element of (a) is of the form $ra + na$ where $r \in R$ and $n \in \mathbb{Z}$. If R has an identity, then each element of (a) can be written as $\sum x_i a y_i, (x_i, y_i \in R)$. If R is a commutative ring with an identity, then $(a) = \{ra : r \in R\}$.

It is easy to see that the intersection of two ideals of a ring R is still an ideal of R. In general, if $\{A_\alpha\}_{\alpha \in B}$ is a nonempty family of ideals of R, then $\cap_{\alpha \in B} A_\alpha$ is an ideal of R. Let T be a nonempty set of R and let $\{A_\alpha\}$ be the set of ideals of R that contains T (such ideals exist, e.g., R is one of them). Like the case for subrings, we call $\cap A_\alpha$ the ideal of R generated by T and denoted by (T). In particular, when $T = \{a\}$, (T) is the principal ideal generated by a. When $T = \{a_1, a_2, \cdots, a_n\}$, (T) is denoted by $(a_1, a_2, \cdots, a_n)$.

It is natural to ask what elements of R are in (T). It is easy to show that

$$(T) = \left\{ \sum x_i | x_i \in (t_i), t_i \in T \right\}.$$

Example 9.29 *If $\mathbb{Z}[x]$ is the polynomial ring of one variable over the ring of integers $\mathbb{Z}$, then the ideal $(2, x)$ is not a principal ideal.*

Proof Since $\mathbb{Z}[x]$ is a commutative ring with the identity, so

$$(2, x) = \{2f(x) + xg(x) | f(x), g(x) \in \mathbb{Z}[x]\},$$

and hence,

$$(2, x) = \{2a_0 + a_1 x + \cdots + a_n x^n | a_i \in \mathbb{Z}, n \geq 0\}.$$

If $(2, x)$ is a principal ideal, say $(2, x) = (p(x))$, then

$$2 = q(x)p(x), \quad x = h(x)p(x).$$

From $2 = q(x)p(x)$, we know that $p(x) = a$ and $a = \pm 1$ or ± 2. From $x = ah(x)$, we know that $a = \pm 1$. Therefore, $\pm 1 = p(x) \in (2, x)$ and $(2, x) = \mathbb{Z}[x]$. This is a contradiction, and the result is proved.

Using ideals, we can define the concept of quotient rings. Given a ring R and an ideal A, with respect to addition, R is an additive group and A is an invariant subgroup of R. Therefore, we naturally have an addition on the set

$$\frac{R}{A} = \{\bar{a} | a \in R\}$$

of the cosets of A:

$$\bar{a} + \bar{b} = \overline{a + b}.$$

where $\bar{a} = \{a + x | a \in R, x \in A\}$. R/A is an additive group with respect to the addition we just described, and we usually call R/A the residue classes modulo A. Now we define the multiplication operation on R/A

$$\bar{a} \cdot \bar{b} = \overline{ab}.$$

It is straightforward to verify that $(R/A, +, \cdot)$ is a ring.

Definition 9.30 Let R be a ring and A an ideal of R. The ring R/A is called the quotient ring of R with respect to A. The quotient ring R/A is also called the ring of residue classes of R modulo A. □

Example 9.31 *Consider the ring of integers $\mathbb{Z}$ and its principal ideal (m), the quotient ring $\mathbb{Z}/(m)$ contains m elements. Any element $\bar{a}$ consists of integers that have remainder a dividing by m. Thus, $\bar{a}$ is called a residue class modulo m, and $\mathbb{Z}/(m)$ is called the ring of residue classes modulo m.*

Example 9.32 *The set $F[x]$ of all polynomials with coefficients from a field of numbers F is a ring with respect to the addition and multiplication of polynomials and is called a polynomial ring. For a polynomial $f(x)$ over F with degree n, the ideal generated by $f(x)$ is*

$$(f(x)) = \{f(x)g(x) | g(x) \in F[x]\},$$

and the corresponding quotient ring is

$$\frac{F[x]}{(f(x))} = \{\overline{r(x)} | r(x) \in F[x], \text{ the degree of } r(x) \text{ is less than } n\}.$$

We can further prove that $F[x]/(f(x))$ is a field if and only if $f(x)$ is an irreducible polynomial.

Proof The first portion is easy, so we will only prove the last part. Let $f(x)$ be an irreducible polynomial. If $\overline{g(x)} \neq 0$, then $f(x) \nmid g(x)$. Since $f(x)$ is irreducible, $g(x)$ and $f(x)$ must be coprime. Thus, there are polynomials $s(x)$ and $t(x)$ such that

$$s(x)f(x) + t(x)g(x) = 1,$$

and hence,

$$\overline{t(x)} \cdot \overline{g(x)} = \overline{1}.$$

This means that $\overline{g(x)}$ has an inverse $\overline{t(x)}$ and the sufficiency is proved.

Conversely, if $f(x)$ is reducible, then there are two polynomials $g(x)$, $h(x)$ of degree less than n such that $f(x) = g(x)h(x)$. Therefore,

$$\overline{h(x)} \cdot \overline{g(x)} = \overline{0}.$$

This means that $F[x]/(f(x))$ has zero divisors and we get a contradiction. The necessity is proved.

Definition 9.33 Let R, R' be two rings. If there is a map $f : R \to R'$, such that

$$f(a + b) = f(a) + f(b), \quad f(ab) = f(a) \cdot f(b),$$

hold for all $a, b \in R$, then f is said to be a homomorphism from R to R'. If f is a surjection from R to R', then f is said to be a surjective homomorphism and denoted by $R \sim R'$. If the homomorphism f from R to R' is an injection, then f is said to be an injective homomorphism. If f is a bijection from R to R', then f is said to be an isomorphism from R to R'. In this case, R and R' are said to be isomorphic, and this is denoted by $R \cong R'$. □

We have the following properties for rings and surjective homomorphisms.

Theorem 9.34 *Let R and R' be two rings, and suppose there is a surjective homomorphism from R to R'. Then the image of the zero of R is zero of R', the negative of element a from R is the negative of the image of a. If R is commutative, then so is R'. If R has the identity 1, then R' also has identity $\overline{1}$, and the image of the identity is the identity.*

Theorem 9.35 *Let R and R' be two rings and f a surjective homomorphism from R to R'. Let A be an ideal of R and $A \subset \ker f$, then there is a unique surjective homomorphism f_* from R/A to R' such that $f = f_* \circ \varphi$, where φ is the natural homomorphism from R to R/A (i.e., the surjective homomorphism from R to R/A that maps element a to the coset $a + A$). $\ker f = A$ if and only if f_* is an isomorphism from R/A to R'.*

Proof Since f, φ are homomorphisms for additive groups, by the fundamental homomorphism theorem, there is a unique f_* (as a group homomorphism) with the required property. If $x \in R$, then

$$f(x) = (f_* \circ \varphi)(x) = f_*(\varphi(x)).$$

Thus, for $a, b \in R$,

$$f_*(\varphi(a)\varphi(b)) = f_*(\varphi(ab)) = f(ab) = f(a)f(b) = f_*(\varphi(a))f_*(\varphi(b)).$$

This shows that f_* preserves multiplication, and hence, f_* is a ring homomorphism. The theorem is proved.

By this theorem, we get the following.

Theorem 9.36 (*Homomorphism theorem for rings*) *Let R be a ring. Then any quotient ring of R is an image of R under a homomorphism. Conversely, if R' is the image of R under the homomorphism f, then $R' \cong R/\ker f$.*

9.4 CHINESE REMAINDER THEOREM

The Chinese remainder theorem not only has important applications in cryptographic analysis and design, but also it is an important tool in polynomial factorization and solving congruence equations.

Definition 9.37 Let R be a commutative ring with identity.

1. If $P \neq R$ is an ideal of R and for any $a, b \in R$, $ab \in P$ implies that $a \in P$ or $b \in P$, then P is called a prime ideal of R.

2. If $M \neq R$ is an ideal of R and for any ideal $I \subset R$, $M \subset I$ implies that $I = M$ or $I = R$, then M is called a maximal ideal of R. □

Definition 9.38 Let R be a commutative ring with identity. If two ideals I, J of R satisfy $I + J = R$, where $I + J = \{a + b | a \in I, \ b \in J\}$, then the ideals I, J are said to be coprime. □

Theorem 9.39 *Let R be a commutative ring with identity and $P \subset R$ is an ideal of R. Then P is a prime ideal if and only if R/P is an integral domain.*

Proof If two elements $\overline{a}, \overline{b}$ of R/P satisfy $\overline{a}\overline{b} = 0$, then $ab \in P$. Since P is a prime ideal, $ab \in P$ implies that $a \in P$ or $b \in P$. This shows $\overline{a} = 0$ or $\overline{b} = 0$, and hence, R/P is an integral domain.

Conversely, if $ab \in P$, then $\overline{ab} = \overline{a}\overline{b} = 0$. As R/P is an integral domain, so $\overline{a} = 0$ or $\overline{b} = 0$, that is, $a \in P$ or $b \in P$, and hence, P is a prime ideal.

Theorem 9.40 *Let R be a commutative ring, and $M \subset R$ is an ideal. Then M is a maximal ideal if and only if R/M is a field.*

Proof Let M be a maximal ideal. Suppose that $\overline{x} \in R/M$ and $\overline{x} \neq 0$, then $x \notin M$ and $M \subsetneq M + xR = R$. Thus, there is an $m \in M$, $y \in R$ such that $m + xy = 1$, and this implies that $\overline{x} \cdot \overline{y} = \overline{1}$. So R/M is a field.

Conversely, let R/M be a field. Suppose J is an ideal of R and $J \supset M$. If $J \neq M$, then there exists $x \in J$ such that $x \notin M$, that is, $\overline{x} \neq 0$. Since R/M is a field, so there is a $\overline{y} \in R/M$ such that $\overline{x} \cdot \overline{y} = \overline{xy} = 1$. This is equivalent to the existence of $m \in M$ such that $xy = 1 + m$. Because $xy \in J$ and $m \in M \subset J$, so $1 = xy - m \in J$. Therefore, $J = R$, and M is a maximal ideal.

Theorem 9.41 (*Chinese remainder theorem*) *Let R be a commutative ring with identity and $I_1, \cdots, I_k$ be pairwise coprime ideals of R. Then*

$$\frac{R}{\bigcap_{i=1}^{k} I_i} \cong \bigotimes_{i=1}^{k} \frac{R}{I_i},$$

where $\bigotimes_{i=1}^{k} R/I_i$ is the Cartesian product, namely, $\bigotimes_{i=1}^{k} R/I_i = \{(\alpha_1, \cdots, \alpha_k) | \alpha_i \in R/I_i, i = 1, \cdots, k\}$.

Proof Define map:

$$\sigma : R \to \bigotimes_{i=1}^{k} \frac{R}{I_i},$$

$$x \mapsto (x \bmod I_1, x \bmod I_2, \cdots, x \bmod I_k) \triangleq (\overline{x}_1, \overline{x}_2, \cdots, \overline{x}_k).$$

It is easy to see that σ is a homomorphism.

We first prove that $ker(\sigma) = \bigcap_{i=1}^{k} I_i$. In fact, for any $x \in ker(\sigma)$,

$$\sigma(x) = (0, \cdots, 0).$$

This yields $x \bmod I_i = 0,\ i = 1, \cdots, k$, and hence, $x \in \bigcap_{i=1}^{k} I_i$.

Conversely, $x \bmod I_i = 0,\ i = 1, \cdots, k$ for any $x \in \bigcap_{i=1}^{k} I_i$, so

$$\sigma(x) = (x \bmod I_1, x \bmod I_2, \cdots, x \bmod I_k) = (0, \cdots, 0).$$

Therefore, $x \in ker(\sigma)$.

Next, we prove that σ is a surjection. To this end, we need to show that for any $(x_1, \cdots, x_k) \in \bigotimes_{i=1}^{k} R/I_i$ there exists $x \in R$ such that $x \bmod I_i = x_i,\ i = 1, \cdots, k$. Since $I_1, \cdots, I_k$ are pairwise coprime, we can find $a_i \in I_1,\ b_i \in I_i$, such that $a_i + b_i = 1 \in R,\ i = 2, \cdots, k$. Therefore,

$$1 = \prod_{i=2}^{k}(a_i + b_i) = (a_2 + b_2) \cdots (a_k + b_k) \triangleq y + b_2 \cdots b_k,$$

where $b_2 \cdots b_k \in \prod_{i=2}^{k} I_i$ and $y \in I_1$. Let $y_1 = b_2 \cdots b_k$, then

$$\begin{cases} y_1 \equiv 1 \bmod I_1, \\ y_1 \equiv 0 \bmod \left(\prod_{i=2}^{k} I_i \right). \end{cases}$$

Similarly, we can find $y_2, \cdots, y_k$ such that

$$\begin{cases} y_i \equiv 1 \bmod I_i, \\ y_i \equiv 0 \bmod \left(\prod_{j=1, j \neq i}^{k} I_j \right). \end{cases}$$

Set $x = \sum_{i=1}^{k} x_i y_i$, then x is what we wanted.

The Chinese remainder theorem we covered in elementary number theory is in fact a special case of Theorem 9.41. Next, we present a useful corollary of the Chinese remainder theorem—the Lagrange interpolation formula. This formula has been used in the construction of trap-door cryptosystems.

Corollary 9.42 (*Lagrange Interpolation Formula*) *Let* $\mathbb{F}$ *be filed. Given* n *points* $(u_1, v_1), \cdots, (u_n, v_n)$, *where* $u_1, \cdots, u_n \in \mathbb{F}$ *are distinct numbers, and let* $g(x) \in \mathbb{F}(x)$ *be a polynomial of degree at most* $n - 1$ *that passes through these* n *points. Then*

$$g(x) = \sum_{i=1}^{n} v_i \prod_{\substack{1 \le j \le n \\ j \ne i}} \frac{x - u_j}{u_i - u_j}.$$

Proof The main idea of proving the formula is to solve the following system of equations using the Chinese remainder theorem:

$$g(x) \equiv v_i (\mathrm{mod}\,(x - u_i)), \ 1 \le i \le n.$$

9.5 EUCLIDEAN RINGS

This section will introduce an important class of rings—the Euclidean rings. In a Euclidean ring, we can define the greatest common divisors as we did in the ring of integers. A Euclidean ring is a unique factorization domain. Let us start with the definition of Euclidean rings.

Definition 9.43 Let R be a ring. If there is a map $\phi : R^* \to \mathbb{N} \bigcup \{0\}$ such that for any $a \in R^*$, $b \in R$, there are $q, r \in R$ that satisfy

$$b = qa + r,$$

where $r = 0$ or $\phi(r) < \phi(a)$, then R is said to be a Euclidean ring. □

It is easy to see that the ring of integers is a Euclidean ring.

Before getting into the properties of Euclidean rings, we define the concept of principal ideal domain.

Definition 9.44 An integral domain R is called a principal ideal domain, if every ideal of R is a principal ideal. □

Theorem 9.45 *If an integral domain R is an Euclidean ring, then R must be a principal ideal domain.*

Proof Let A be an ideal of R. If A contains only zero, then A is a principal ideal. Assume that A has an element other than zero. By the definition of Euclidean rings, there is a map ϕ defined on all nonzero elements of A. The set

$$S = \{\phi(x)|\phi(x) > 0, \ x \in A\}$$

is nonempty and has a smallest positive integer, say $\phi(x_0)$ for some $x_0 \in A$. We will prove $A = (x_0)$. $(x_0) \subseteq A$ is obvious. We only need to prove $A \subseteq (x_0)$. As R is an Euclidean ring, given $a \in A$, there exists $q, r \in R$ such that

$$a = qx_0 + r,$$

where $r = 0$ or $\phi(r) < \phi(x_0)$. By $r \in A$ and the minimality of $\phi(x_0)$, we know that $r = 0$. So $a \in (x_0)$, and $A \subseteq (x_0)$. The theorem is proved.

Example 9.46 *The polynomial ring $\mathbb{F}[x]$ over a field $\mathbb{F}$ is a Euclidean ring.*

Proof Define map from $\phi : \mathbb{F}[x] \to \mathbb{N} \cup \{0\}$ by $\phi(f(x)) = \deg f(x)$. Let $g(x) \in F[x], g(x) \neq 0$. Then for any $f(x) \in \mathbb{F}[x]$, there exist $q(x), r(x) \in \mathbb{F}[x]$ such that

$$f(x) = q(x)g(x) + r(x),$$

where $r(x) = 0$ or $\deg r(x) < \deg g(x)$. So $\mathbb{F}[x]$ is a Euclidean ring.

Example 9.47 *Gaussian integral domain R, which is an integral domain formed by complex numbers of the form $a + bi$ (a, b are arbitrary integers), is an Euclidean ring.*

Proof For any $\alpha = a + bi \in R^*$, define $\phi(\alpha) = a^2 + b^2$. Then ϕ is a map from R^* to $\mathbb{N} \cup \{0\}$. For $\alpha = a + bi, \beta = c + di \in R^*$, a simple calculation shows that

$$\phi(\alpha\beta) = \phi(\alpha)\phi(\beta).$$

Write $\alpha^{-1}\beta = k + li$, where k, l are rational numbers. Take k', l' to be the closest integers to k, l, respectively, that is,

$$|k - k'| \leq \frac{1}{2}, |l - l'| \leq \frac{1}{2}.$$

Let $\gamma = k' + l'i$, then

$$\phi(\alpha^{-1}\beta - \gamma) = (k - k')^2 + (l - l')^2 \leq \frac{1}{4} + \frac{1}{4} = \frac{1}{2}.$$

Let $\delta = \beta - \alpha\gamma$, then $\beta = \alpha\gamma + \delta$. If $\delta \neq 0$, then

$$\phi(\delta) = \phi(\beta - \alpha\gamma) = \phi(\alpha(\alpha^{-1}\beta - \gamma)) = \phi(\alpha)\phi(\alpha^{-1}\beta - \gamma)$$
$$\leq \frac{1}{2}\phi(\alpha) < \phi(\alpha).$$

The result is proved.

9.6 FINITE FIELDS

Finite fields have been widely used in cryptography and coding theory. In this section, we will mainly discuss the structure of finite fields and some special examples.

Definition 9.48 A field that contains only finitely many elements is called a finite field. ☐

Definition 9.49 A field that does not contain a proper subfield is called a prime field. ☐

A prime field of characteristic p is obviously a finite field. It is easy to see that this field is the complete system of residues modulo p. We denote it as $\mathbb{F}_p$.

Theorem 9.50 *For any prime p and any positive integer d, there is a finite field that has p^d elements. This finite field is denoted as $\mathbb{F}_{p^d}$.*

Proof Consider the polynomial $f(x) = x^{p^d} - x$ over the field $\mathbb{F}_p$. It can be seen easily that $f(\alpha) = 0$ for any $\alpha \in \mathbb{F}_p$. Let S be the set roots of $f(x)$ and F is the smallest field that contains S. It can be seen easily that F has characteristic p and $\mathbb{F}_p \subset F$. For any $\alpha, \beta \in S$,

$$f(\alpha - \beta) = (\alpha - \beta)^{p^d} - (\alpha - \beta) = (\alpha^{p^d} - \alpha) - (\beta^{p^d} - \beta) = f(\alpha) - f(\beta) = 0,$$
$$f(\alpha\beta^{-1}) = (\alpha\beta^{-1})^{p^d} - (\alpha\beta^{-1}) = \beta^{-p^d}f(\alpha) + \alpha f(\beta^{-1}) = 0.$$

These mean that S is a field, that is, $S = F$. It is obvious that S has p^d elements.

Remark Let $g(x)$ be an irreducible polynomial of degree d over $\mathbb{F}_p[x]$. Then $\mathbb{F}_p[x]/(g(x))$ is a finite field with p^d elements. Finite fields of p^d elements are unique up to isomorphism.

Theorem 9.51 *Let $\mathbb{F}_q$ be a finite field of characteristic p. Then the set $\mathbb{F}_q^*$ of nonzero elements of the field forms a cyclic group with respect to the field multiplication.*

Proof Write $q - 1 = p_1^{\alpha_1} \cdots p_s^{\alpha_s}$. By Theorem 8.28, we know that $a^{q-1} - 1 = 0$ holds for any $a \in \mathbb{F}_q^*$. Consider the polynomial $f_i(x) = x^{p_i^{\alpha_i}} - 1$ over $\mathbb{F}_q$. Since $f_i(x)|x^{q-1} - 1$, so all the roots of $f_i(x)$ are contained in $\mathbb{F}_q$ and the number of such roots is $p_i^{\alpha_i}$. Now consider the polynomial $h_i(x) = x^{p_i^{\alpha_i - 1}} - 1$ over $\mathbb{F}_q$. Since $h_i(x)|f_i(x)$, each root of $h_i(x)$ is a zero of $f_i(x)$, and we have $p_i^{\alpha_i - 1}$ such roots. This implies that there is an $a_i \in \mathbb{F}_q$, such that $f_i(a_i) = 0$, but $h_i(a_i) \neq 0$, that is, the order of $a_i \in \mathbb{F}_q^*$ is $p_i^{\alpha_i}$. Let $a = a_1 \cdots a_s$. Then the order of a is $q - 1$, so $\mathbb{F}_q^*$ is a cyclic group generated by a.

Theorem 9.52 *The finite field $\mathbb{F}_{p^n}$ contains the finite field $\mathbb{F}_{p^m}$ if and only if $m|n$.*

Proof Since $p^m - 1|p^n - 1$ if and only if $m|n$, so the finite field $\mathbb{F}_{p^n}$ contains the finite field $\mathbb{F}_{p^m}$ if and only if $m|n$.

Theorem 9.53 *Let n be a positive integer. Then the polynomial $x^{q^n} - x$ over $\mathbb{F}_q[x]$ is the product of all monic irreducible polynomials over $\mathbb{F}_q[x]$ with degrees divides n.*

Proof It is easy to see that $x^{q^n} - x$ has no roots with multiplicity more than 1. Assume a polynomial $f(x)|x^{q^n} - x$ and $\deg f(x) = d$. By Theorem 9.50, we know that all roots of $f(x)$ are contained in $\mathbb{F}_{q^n}$. From the remark of Theorem 9.50, the set of roots of $f(x)$ generates $\mathbb{F}_q[x]/(f(x))$. Therefore, $\mathbb{F}_q[x]/(f(x)) = \mathbb{F}_{q^d} \subset \mathbb{F}_{q^n}$. By Theorem 9.52, we get $d|n$.

Let $f(x) \in \mathbb{F}_q[x]$ be irreducible with $\deg f(x) = d$ and $d|n$. This implies that $x^{q^d - 1} - 1|x^{q^n - 1} - 1$, and hence, $x^{q^d} - x|x^{q^n} - x$. For any root a of $f(x)$, we have $a^{q^d} - a = 0$, so $x - a|x^{q^d} - x|x^{q^n} - x$. In particular,

$$\gcd(f(x), x^{q^n} - x) \neq 1.$$

Since $f(x)$ is irreducible over $\mathbb{F}_q[x]$, we have $f(x)|x^{q^n} - x$.

9.7 FIELD OF FRACTIONS

It is known that the ring of integers is a subring of the field of rational numbers, and the field of rational numbers is the smallest field that contains the ring of integers. The question we are asking now is for a ring R, whether we can find a field that contains R. In order for the ring R to be contained in a field, a necessary condition is that R does not have zero divisors, furthermore, R must be commutative. In this section, we are able to prove that when R is a commutative ring without zero divisors, then there must be a smallest field such any element in R has an inverse in the field.

Theorem 9.54 *Every commutative ring R without zero divisors is a subring of a field Q.*

Proof If R contains only zero, the theorem obviously holds. Assume that R has at least two elements. We use $a, b, c \cdots$ to denote elements of R and form a set

$$A = \{(a, b) | a, b \in R, b \neq 0\}.$$

A is actually a subset of the Cartesian product $R \times R$. Define a relation on A: $(a, b) \sim (a', b')$ if and only if $ab' = a'b$. This relation obviously satisfies

1. $(a, b) \sim (a, b)$ holds for any $(a, b) \in A$.

2. If $(a, b) \sim (a', b')$, then $(a', b') \sim (a, b)$.

3. If $(a, b) \sim (a', b'), (a', b') \sim (a'', b'')$, then $(a, b) \sim (a'', b'')$.

In other words, $\sim$ is an equivalence relation.

   The equivalence relation partitions A into equivalent classes $\overline{(a, b)}$. We write the equivalent class $\overline{(a, b)}$ as $\frac{a}{b}$, and set

$$Q_0 = \left\{\frac{\overline{a}}{b} | a, b \in R, b \neq 0\right\}.$$

We will show the following define two operations on Q_0:

$$\frac{\overline{a}}{b} + \frac{\overline{c}}{d} = \frac{\overline{ad + bc}}{bd}, \quad \frac{\overline{a}}{b} \cdot \frac{\overline{c}}{d} = \frac{\overline{ac}}{bd} \qquad (*).$$

To check that these are binary operations, we note

1. From $b \neq 0, d \neq 0$, we get $bd \neq 0$ and $\overline{\dfrac{ad+bc}{bd}}$, $\overline{\dfrac{ac}{bd}} \in Q_0$.

2. If $\dfrac{\overline{a}}{b} = \dfrac{\overline{a'}}{b'}$, $\dfrac{\overline{c}}{d} = \dfrac{\overline{c'}}{d'}$, then $ab' = a'b, cd' = c'd$, and hence,

$$(ad+bc)b'd' = (a'd'+b'c')bd, \quad ab'cd' = a'bc'd.$$

This means that

$$\dfrac{\overline{a}}{b} + \dfrac{\overline{c}}{d} = \dfrac{\overline{a'}}{b'} + \dfrac{\overline{c'}}{d'}; \qquad \dfrac{\overline{a}}{b} \cdot \dfrac{\overline{c}}{d} = \dfrac{\overline{a'}}{b'} \cdot \dfrac{\overline{c'}}{d'}.$$

That is, the results of adding and multiplying two classes are independent of the choices of the representatives; hence, they are binary operations.

It is not hard to prove that Q_0 forms a field with respect to the addition and multiplication defined above. Obviously, $\dfrac{\overline{a}}{a}$ is the multiplicative identity, and the multiplicative inverse of $\dfrac{\overline{a}}{b}$ is $\dfrac{\overline{b}}{a}$.

Let R_0 be the subset of Q_0 that consists of elements of the form $\dfrac{\overline{qa}}{q}$ (q is a fixed element and a is an arbitrary element), then the map

$$a \rightarrow \dfrac{\overline{qa}}{q}$$

is actually a bijection from R to R_0. This map is in fact an isomorphism since

$$\dfrac{\overline{qa}}{q} \dfrac{\overline{qb}}{q} = \dfrac{\overline{q^2(ab)}}{q^2} = \dfrac{\overline{q(ab)}}{q}, \quad \dfrac{\overline{qa}}{q} + \dfrac{\overline{qb}}{q} = \dfrac{\overline{q(a+b)}}{q}.$$

Therefore, by the homomorphism theorem of rings, we have proved the existence of a field that contains R.

Since R is included in the field Q, a nonzero element $b \in R$ has an inverse $b^{-1} \in Q$. Furthermore,

$$ab^{-1} = b^{-1}a = \dfrac{a}{b}(a, b \in R, b \neq 0)$$

is meaningful in Q. We have the following.

Theorem 9.55 Q *consists of exactly elements of the form* $\dfrac{a}{b} (a, b \in R,$ $b \neq 0$), *where*

$$\frac{a}{b} = ab^{-1} = b^{-1}a.$$

Proof To prove every element of Q can be written as $\dfrac{a}{b}$, it is sufficient to show that every element of Q_0 is of the form

$$\frac{\overline{\frac{qa}{q}}}{\frac{qb}{q}} = \overline{\frac{qa}{q}} \cdot \overline{\frac{qb}{q}}^{-1}.$$

Consider an arbitrary element $\dfrac{\overline{a}}{b}$ of Q_0. Since

$$\left(\overline{\frac{qb}{q}}\right)^{-1} = \overline{\frac{q}{qb}},$$

we have

$$\overline{\frac{qa}{q}} \left(\overline{\frac{qb}{q}}\right)^{-1} = \overline{\frac{q^2a}{q^2b}} = \overline{\frac{a}{b}} = \frac{\overline{\frac{qa}{q}}}{\overline{\frac{qb}{q}}}.$$

It is obvious that every $\dfrac{a}{b}$ is in Q. The proof is complete.

Since the elements of Q are of the form $\dfrac{a}{b}$, the relation between Q and R is similar to that between the ring of integers and the field of rational numbers.

Definition 9.56 A field Q is called a field of fractions of a ring R, if R is included in Q and Q consists of exactly elements of the form $\dfrac{a}{b}$ $(a, b \in R, b \neq 0)$. □

By Theorems 9.54 and 9.55, we know that a commutative ring without zero divisors that contains at least two elements has at least a field of fractions. In general, a ring may have more than one field of fractions.

Theorem 9.57 *Let R be a commutative ring without zero divisors that has at least two elements and F a field that includes R. Then F includes a field of fractions of R.*

Proof In F

$$ab^{-1} = b^{-1}a = \frac{a}{b}, \quad a, b \in R, b \neq 0,$$

is well defined. So the subset of F

$$\overline{Q} = \left\{ \frac{a}{b} | a, b \in R, b \neq 0) \right\}$$

is obviously a field of fractions of R. The proof is complete.

Note that every field of fractions of R satisfies the operation rules $(*)$. However, $(*)$ is completely determined by the addition and multiplication of R. In other words, the construction of fields of fractions of R is completely determined by the structure of R. Therefore, we have the following.

Theorem 9.58 *The fields of fractions of isomorphic rings are isomorphic.*

EXERCISES

9.1 Prove that $\mathbb{Z}[i] = \{a + bi | a, b \in \mathbb{Z}\}$ is a ring with respect to the addition and multiplication of complex numbers.

9.2 Find all roots of the equation $x^2 - 1 = 0$ in $\mathbb{Z}_{15}$.

9.3 Let A be the set of all reduced fractions with denominators being nonnegative powers of 2. Is A a ring with respect to the addition and multiplication of numbers?

9.4 If the additive group $(R, +)$ of a ring R is cyclic, then R is commutative.

9.5 Let R be a commutative ring, and let a be an ideal of R and S a subset of R. Let

$$(a : S) = \{x | x \in R, xS \subseteq a\}.$$

Prove that $(a : S)$ is an ideal of R.

9.6 Let F be a field. Which elements of the polynomial ring are in the principal ideal (x^2)? What are the elements in $F[x]/(x^2)$?

9.7 Let R be a finite commutative ring with identity. Prove that every element of R is either invertible or a zero divisor. Then use this to prove that a finite integral domain is a field.

9.8 Let A be the ring of even integers and $a = \{4x | x \in \mathbb{Z}\}$. Prove that a is an ideal of A. What kind of ring is A/a? Is a the ideal (4)? Is $A/(4)$ a field?

9.9 Prove that $(3)/(6)$ is an ideal of $\mathbb{Z}/(6)$ and

$$\mathbb{Z}/(6)/(3)/(6) \cong \mathbb{Z}/(3).$$

9.10 Let R be a ring. For every polynomial $g(x) = a_0 + a_1 x + \cdots + a_n x^n \in R[x]$, define $f : R[x] \to R$ by $(g(x)) = a_0$. Prove that f is a surjective homomorphism from $R[x]$ to R. Find $\ker f$. What kind of ring is $R[x]/\ker f$?

9.11 Prove that the ring $\mathbb{Z}[i]$ of Gaussian integers is isomorphic to $\mathbb{Z}[x]/(x^2 + 1)$.

9.12 Find all homomorphisms from $\mathbb{Z}$ to itself and determine their kernels.

CHAPTER 10

Some Mathematical Problems in Public Key Cryptography

CRYPTOGRAPHY IS an applied science that is a key mechanism for ensuring information security. The knowledge of cryptography falls into multiple areas, such as mathematics and computer science. Cryptography, which can be regarded as a subject that has its own theoretical system and characteristics, is mainly based on mathematics as its theoretical foundation and computer science as its source of implementation tools. The main purpose of this chapter is to present a comprehensive introduction to some problems from number theory and algebra that are used in cryptography. These problems constitute the core materials for the mathematical theory that public key cryptography are based upon. Deeper understanding of these problems is very useful to the future study of the techniques of analysis and design of public key cryptography. In addition, the importance of the mathematical problems discussed in this chapter is mainly in their computational efficiency or computational complexity. So issues of computational complexity for some of the problems will also be addressed.

10.1 TIME ESTIMATION AND COMPLEXITY OF ALGORITHMS

First, we need to introduce some concepts from complexity theory; this is helpful in understanding the assessment of security of cryptographic algorithms. Usually, we have two measures of complexity: time complexity and space complexity. These two complexities are functions of the length n of inputs. To some extent, we can exchange computing time for space consumption or vice versa. For example, assume that we are factoring a large integer by trial division. We can use a single computer to perform serial search on the prime table; or we can distribute the prime table to multiple computers to work in parallel. The former sacrifices time while the latter sacrifices space.

Definition 10.1 In the computing of binary integers, a single addition, subtraction, or multiplication of two bits is called a bit operation. □

In addition to the above bit operation, executing an algorithm involves a shifting operation, and a save and fetch operation. These are fairly fast operations and hence are usually neglected in estimating the time complexity of an algorithm.

In an execution of an algorithm, the number of bit operations is proportional to the time spent on the computation. Therefore, the time estimation of an algorithm always means the number of bit operations required by a computer to run the algorithm.

Theorem 10.2 *If a, b are two integers with length k in binary representation, then computing the sum or difference of these integers requires $O(k)$ bit operations.*

Proof Let $a = \sum_{0 \leq i \leq k-1} a_i 2^i$, $b = \sum_{0 \leq i \leq k-1} b_i 2^i$, where $a_i, b_i \in \{0, 1\}$, $i = 0, 1 \cdots, k-1$. Then the following procedure outputs $c = a + b$.

> 1. $\gamma_0 \leftarrow 0$
> 2. for $i = 0, \cdots, k-1$ do
> > $c_i \leftarrow a_i + b_i + \gamma_i$, $\gamma_{i+1} \leftarrow 0$
> > if $c_i \geq 2$ then
> > > $c_i \leftarrow c_i - 2$, $\gamma_{i+1} \leftarrow 1$
> 3. $c_k \leftarrow \gamma_k$
> 4. return $c = \sum_{0 \leq i \leq k} c_i 2^i$

It can be seen that computing $a + b$ requires at most $2k$ bit operations.

Theorem 10.3 *If a and b are two integers with lengths k and l, respectively, and $k \leq l$, then computing the product or quotient of these integers requires $O(l^2)$ bit operations.*

Proof Let us check the implementation of multiplication first. Let $a = \sum_{0 \leq i \leq k-1} a_i 2^i$, $b = \sum_{0 \leq j \leq l-1} b_j 2^j$, where $a_i, b_j \in \{0,1\}$, $i = 0, 1 \cdots, k-1$, $j = 0, 1 \cdots, l-1$. Then the following procedure outputs $c = ab$.

> 1. for $i = 0, \cdots, k-1$ do
> $$d_i \leftarrow a_i 2^i b$$
> 2. return $c = \sum_{0 \leq i \leq k-1} d_i$

It can be seen that the multiplication can be divided into two steps:

(i) For each $a_i (0 \leq i \leq k-1)$, compute $a_i \sum_{0 \leq j \leq l-1} b_j 2^j$. Since a_i takes value of 0 or 1, the value of $a_i \sum_{0 \leq j \leq l-1} b_j 2^j$ is either 0 or $\sum_{0 \leq j \leq l-1} a_i b_j 2^j$. This requires at most l bit operations.

(ii) Use $a_i \sum_{0 \leq j \leq l-1} b_j 2^j$ to multiply 2^i, sum them over $i = 1, 2, \cdots, k-1$, then the product ab is obtained. Multiplication by 2^i is just a shifting operation. Then the summation is on k numbers each of which is at most $k+l$ bits. Therefore, the number of bit operations for completing the multiplication is $O(k(l+k))$. Since $k \leq l$, $O(k(l+k)) = O(kl)$.

The proof for division is similar and is left to the reader as an exercise.

Remark The time complexity of multiplication of two n-bit integers is $O(n^{1.59})$ if the Karatsuba algorithm is used. The complexity can be $O(n \log n \log \log n)$ if the fast Fourier transform is used.

By Theorem 10.3, we can prove the following.

Corollary 10.4 *The number of bit operations for computing $n!$ is $O(n^2 \log_2^2 n)$.*

Next, we consider the time estimation of the Euclidean algorithm which we introduced in elementary number theory.

Theorem 10.5 *Let a, b be two integers with the binary lengths n and m, respectively. Assume $n \geq m$. Then the time complexity for computing (a, b) by using the Euclidean algorithm is $O(\log_2^2 a)$.*

Proof For the given integers a, b, we have

$$
\begin{aligned}
a &= q_0 b + r_0, & 0 < r_0 < |b|, \\
b &= q_1 r_0 + r_1, & 0 < r_1 < r_0, \\
r_0 &= q_2 r_1 + r_2, & 0 < r_2 < r_1, \\
&\cdots \quad \cdots \quad \cdots & \cdots \quad \cdots \quad \cdots \\
r_{k-5} &= q_{k-3} r_{k-4} + r_{k-3}, & 0 < r_{k-3} < r_{k-4}, \\
r_{k-4} &= q_{k-2} r_{k-3} + r_{k-2}, & 0 < r_{k-2} < r_{k-3}, \\
r_{k-3} &= q_{k-1} r_{k-2}.
\end{aligned}
$$

This can be turned to the following pseudo code:

```
1. r₀ ← a,  r₁ ← b
2. i ← 1
     while rᵢ ≠ 0 do
          rᵢ₊₁ ← rᵢ₋₁rem rᵢ, i ← i+1
3. return rᵢ₋₁.
```

Here $r_{i-1} \mathrm{rem}\, r_i$ stands for the remainder of r_{i-1} divided by r_i.

The following facts are needed for estimating the complexity of the algorithm:

(i) The algorithm needs at most $2n$ divisions.

(ii) The time complexity of the division (with remainder) operation $a = bq + r$ is $O((\log_2 a)(\log_2 q))$.

(iii) The quotients $q_0, q_2, \cdots, q_{k-1}$ in the above algorithm sanctifies

$$
\sum_{i=0}^{k-1} \log_2 q_i = \log_2 \prod_{i=0}^{k-1} q_i \leq \log_2 a.
$$

Therefore, the total number of operations for Euclid's algorithm becomes

$$
O\left((\log_2 a) \sum_{i=0}^{k-1} \log_2 q_i \right) = O\left((\log_2 a)^2 \right).
$$

This proves that the time complexity for computing (a, b) is $O(\log_2^2 a)$.

The (extended) Euclidean algorithm is needed in problems such as finding inverse modulo m and solving modular linear equations of one variable. We can get the following corollaries of Theorem 10.5.

Corollary 10.6 *Given a modulus m, for any a with $(a, m) = 1$, the time for computing a^{-1} (mod m) by using the (extended) Euclidean algorithm is $O(\log_2^2 m)$.*

Corollary 10.7 *Given a modulus m, for any a, b with $(a, m)|b$, the time for solving $ax \equiv b$ (mod m) by using the (extended) Euclidean algorithm is $O(\log_2^2 m)$.*

These corollaries use the following extended Euclidean algorithm for finding x, y such that $xa + yb = (a, b)$:

1. $r_0 \leftarrow a,\quad s_0 \leftarrow 1,\quad t_0 \leftarrow 0,$
   $r_1 \leftarrow b,\quad s_1 \leftarrow 0,\quad t_1 \leftarrow 1$
2. $i \leftarrow 1$
   while $(r_i \neq 0)$ do
   $\quad\quad q_i \leftarrow r_{i-1} \text{quo } r_i$
   $\quad\quad r_{i+1} \leftarrow r_{i-1} \text{rem } r_i$
   $\quad\quad s_{i+1} \leftarrow s_{i-1} - q_i s_i$
   $\quad\quad t_{i+1} \leftarrow t_{i-1} - q_i t_i$
   $\quad\quad i \leftarrow i+1$
3. $l \leftarrow i - 1$
4. return l, r_i, s_i, t_i for $0 \leq i \leq l+1$, and q_i for $1 \leq i \leq l$

In this algorithm, r_{i-1}quo r_i stands for the quotient of r_{i-1} dividing by r_i. This algorithm yields

(i) For any $0 \leq i \leq l$, $(a, b) = (r_i, r_{i+1}) = r_l$.

(ii) For any $0 \leq i \leq l$, $s_i a + t_i b = r_i$. In particular, we have $xa + yb = (a, b)$ by setting $x = s_l, y = t_l$.

The operation of modulo of powers is a common operation in many public key algorithms, especially for public key systems based on integer factorization and discrete logarithm. Before giving the estimation of modulo of powers, we present a widely used algorithm for modulo of powers—the modulo square and multiply method.

Given modulus m and $0 \le a < m$, $0 \le x < \varphi(m)$, the following procedure returns $a^x \pmod{m}$. Write $x = \sum_{0 \le i \le k} b_i 2^i$:

1. $y \leftarrow 1$
2. for $i = k, k-1, k-2, \cdots, 0$ do
   $\quad y \leftarrow y^2 a^{b_i} \pmod{m}$
3. return y

We have the following theorem on the complexity of this algorithm.

Theorem 10.8 *Given modulus m and $0 \le a < m$, $0 \le x < \varphi(m)$, the number of bit operations for computing $a^x \pmod{m}$ using the modulo square and multiply method is $O(\log_2^3 m)$.*

Next, we define the class of polynomial-time algorithms.

Definition 10.9 An algorithm is said to be polynomial-time if the number of steps required to return the result for a given input is $O(k^d)$ for some nonnegative integer d, where k is the length of input. □

In fact, checking whether an algorithm is a polynomial-time algorithm requires finding a polynomial $P(x)$, such that the algorithm can be completed in a number of bit operation that is less than or equal to $P(k)$.

A problem is a question of a general form that needs to be answered. Examples of questions include the factorization problem, the discrete logarithm problem, and the problem of finding the greatest common factors.

Complexity theory divides problems into different classes. The classification is based on the cost of time and space for solving the hardest instance of the problem using a Turing machine. This is a finite state machine with an infinite strip of tape to perform read–write operations. It has a controller of finitely many states, a read–write head, and a tape with infinitely many cells. The read–write head responds to the state the machine is currently in and records on the tape, until the end. This is, in fact, a theoretical model for computers.

Definition 10.10 Let M be a problem. If there is a polynomial $P(x)$ such that for any instance I of size n of M, the number of operations $T_M(n)$ for solving I on a Turing machine satisfies $T_M(n) \le P(n)$, then M is said to belong to the class P. □

Obviously, the operations of adding, subtracting, multiplying, and dividing, the Euclidean algorithm, the method of finding modulo inverse, the algorithm for solving linear modulo equations we discussed before are in the class P.

Now let us modify the Turing machine we described before, to construct a nondeterministic Turing machine (NTM): We add a guessing head that writes the best state guess and validates by a finite states controller. If a decision problem can be verified in finite number of steps on a nondeterministic Turing machine, then this problem is said to be NTM solvable.

Definition 10.11 Let M be a problem. If there is a polynomial $P(x)$, such that for any instance I of M with size n, there is a guess ω for solving I, and the number $T_M(n)$ for validating the guess ω on a nondeterministic Turing machine satisfying $T_M(n) \leq P(n)$, then the problem is said to be in the class NP. □

First, the class P is a subset of the class NP ($P \subseteq NP$). However, the conclusion $NP \subseteq P$ has not yet been proven. Many people believe that there are NP problems do not have a polynomial algorithm, that is, $P \neq NP$, even though no proof is available. Some hard problems in number theory, such as factorization of big integers, are of NP class. The security of several public key algorithms is based on the assumption that $P \neq NP$. That explains the importance of computational complexity theory in cryptography.

Let M_1, M_2 be two problems. If there is a polynomial algorithm that transforms any instance of M_1 to an instance of M_2, then we say that M_1 can be reduced to M_2 in polynomial time, and this is denoted as $M_1 \propto M_2$. Two problems that can be mutually reduced in polynomial time are said to be equivalent in polynomial time. More generally, if two problems M_1, M_2 are equivalent above a secure probability ρ (i.e., the success probability of transforming any instance of M_1 to an instance of M_2 is at least ρ), then M_1, M_2 are said to be equivalent in probabilistic polynomial time.

Definition 10.12 If a decision problem $M \in NP$ and for any other decision problem $M' \in NP$ we have $M' \propto M$, then M is said to belong to class NP-C. □

If there is a decision problem M that belongs to class $NP\text{-}C$ that can be proved to be solvable using a polynomial time algorithm (i.e., $M \in P$), then $NP = P$ holds obviously. Otherwise, $NP\text{-}C \subseteq NP \setminus P$ and all $NP\text{-}C$ problems are hard. This is a result that a cryptographer hopes to have.

The Boolean satisfiability problem (SAT problem) was the first known $NP\text{-}C$ problem. $NP\text{-}C$ also includes the vertex cover problem and the Hamiltonian cycle problem.

10.2   INTEGER FACTORIZATION PROBLEM

Many public key cryptographic algorithms are based on the difficulty of integer factorization. This means that if the problem of integer factorization could be solved in polynomial time or with possible computing resources, then these cryptographic algorithms would be broken. For example, from the encrypted ciphertexts one could recover the corresponding plaintexts; digital signatures could be forged by malicious users; even digital money could be repeatedly used without being traced.

Therefore, for these problems, besides knowing that they are difficult, we also need to learn the fastest algorithms for solving them up to date so that we can determine the lengths of security parameters for using them in public key cryptography. For example, the problem of factoring $n = pq$ has been used in the design of the famous public key encryption algorithm since 1978, where p, q are two large primes of similar lengths. The security length of $n = pq$ has been increased to 2048 bits from the early setup of 256 bits. This reflects the impact of increasing computer speed coupled with the improvement of methods for solving related challenging problems in cryptographic algorithms.

The oldest algorithm of finding factors of big integers is the Sieve of Eratosthenes introduced in elementary number theory, which is also called trial division. The method tries each prime that is not larger than $\sqrt{n}$ and checks whether it divides n. The number of divisions required for this method is $O(\frac{\sqrt{n}}{\ln n})$. There are two fast factorization methods available currently: One is the widely used quadratic sieve algorithm (QS). This algorithm is the fastest method at the moment for numbers that are less than 110 digits in decimal representation whose asymptotic computational time is

$$e^{(1+O(1))(\ln n)^{\frac{1}{2}}(\ln\ln n)^{\frac{1}{2}}}.$$

The other method is the number field sieve (NFS) algorithm, which is the fastest known method for factoring large integers (e.g., more than 110 digits). The asymptotic computational time for NFS is

$$e^{(1.923+O(1))(\ln n)^{\frac{1}{3}}(\ln\ln n)^{\frac{2}{3}}}.$$

Integer factorization is a rapidly developing research area, and the advancement of mathematics theory is unpredictable. Even though the problem is not completely solved yet by many different attempts, some progress has been made. Improved factorization methods together with advances in computer speed will result in further increase of the length of n (the security parameter). In the 1970s, it was hard to factor an integer of more than 100 bits given the computer speed and factorization formula for that time. So the security length of $n = pq$ was set to be 256 bits. In the early 1990s, without even applying the fastest NSF algorithm, an integer of 426 bits was successfully factored by using multiple computers connected over the Internet. It is believed that a modulus of 512 bits can be easily attacked by a collaboration of a large group. The current recommended security parameter is 2048 bits.

RSA Laboratories initiated the RSA Factoring Challenge to keep pace with the development of factorization. It contains a series of numbers that are hard to factorize, from 100 digits to 500 digits. But the challenge ended in 2007. As of December 2009, the length of the RSA challenge being factored was 768 bits.

10.3 PRIMALITY TESTS

Prime numbers are of special importance in cryptography, especially in public key cryptography. For example, in setting up some parameters of certain cryptographic protocols, we need to select large prime numbers. Given a random number, determining whether it is a prime number is called a primality test problem. In 2002, this problem was proved to be a P problem by Indian computer scientists.

The main purpose of this section is to describe two probabilistic algorithms for testing primality.

Lemma 10.13 *Let n be a prime. Then for any $b \in \mathbb{Z}_n^*$, we have*

$$b^{\frac{n-1}{2}} \equiv \left(\frac{b}{n}\right) \pmod{n},$$

where $\left(\dfrac{b}{n}\right)$ is the Jacobi symbol.

Definition 10.14 Let n be an odd composite number and $(b, n) = 1$. If

$$b^{\frac{n-1}{2}} \equiv \left(\frac{b}{n}\right) \pmod{n}, \tag{10.1}$$

then n is called an Euler pseudoprime number to the base b. ☐

Theorem 10.15 *Let n be an odd composite number. Then there are at least half of b's in $\mathbb{Z}_n^*$ such that Equation 10.1 does not hold.*

Proof Let $b_1, \cdots, b_k$ be all the numbers in $\mathbb{Z}_n^*$ that satisfy Equation 10.1. Let $b \in \mathbb{Z}_n^*$ such that Equation 10.1 does not hold (such b exists). By properties of the Jacobi symbol, for $1 \leq i \leq k$, since

$$b^{\frac{n-1}{2}} \not\equiv \left(\frac{b}{n}\right) \pmod{n}, \quad b_i^{\frac{n-1}{2}} \equiv \left(\frac{b_i}{n}\right) \pmod{n},$$

we have

$$\left(\frac{b}{n}\right)\left(\frac{b_i}{n}\right) = \left(\frac{bb_i}{n}\right) \not\equiv (bb_i)^{\frac{n-1}{2}} \pmod{n}.$$

This means that none of $bb_1, \cdots, bb_k$ satisfies Equation 10.1. Note that $bb_1, \cdots, bb_k$ are k distinct elements in $\mathbb{Z}_n^*$, so there are at least k elements in $\mathbb{Z}_n^*$ such that Equation 10.1 does not hold. The result is proved.

Definition 10.16 Let n be an odd composite number and $n - 1 = 2^s t, 2 \nmid t$. Let $b \in \mathbb{Z}_n^*$. If for some r, $0 \leq r < s$,

$$b^t \equiv 1 \pmod{n}, \text{ or } b^{2^r t} \equiv -1 \pmod{n},$$

then n is called a strong pseudoprime number with respect to the base b. □

For strong pseudoprime numbers, we have the following.

Theorem 10.17 *Let n be an odd composite number and $b \in \mathbb{Z}_n^*$. Then the probability that n is a strong pseudoprime with respect to the base b is not greater than $\frac{1}{4}$.*

The proof is complicated and interested readers are referred to the literature [7].

Next, we introduce two methods of primality testing: the Solovay–Strassen algorithm and the Rabin–Miller algorithm.

Solovay–Strassen algorithm

This method uses the Jacobi function to test whether p is a prime number. The steps are as follows:

1. Randomly select a number a that is less than p.

2. Compute the greatest common divisor of a and p. If $(a, p) \neq 1$, then p must be a composite number and the test fails. Return "p is a composite number."

3. Compute $j = a^{\frac{p-1}{2}} \pmod{p}$.

4. Compute the Jacobi symbol $\left(\dfrac{a}{p} \right)$.

5. If $j \neq \left(\dfrac{a}{p} \right)$, then p is not a prime number; if $j = \left(\dfrac{a}{p} \right)$, then p passes the test. From Theorem 10.15, the probability that this number is not prime is not greater than $\dfrac{1}{2}$.

6. If p passes the test, randomly select a new a each time and repeat the above procedure t times. If p passes all the t tests, then the largest probability that p is a composite number is $\dfrac{1}{2^t}$.

We can set the number t depending on the security requirements.

Rabin–Miller algorithm

This is a simple and widely used algorithm. For the number p to be tested, we first compute a positive integer b and an odd integer m such that $p = 1 + 2^b m$. Then perform the following steps:

1. Randomly select a number a that is less than p.

2. Initialize the number of steps $j = 0$ and let $z \equiv a^m \bmod p$.

3. If $z = 1$ or $z = p - 1$, then p is probably a prime, and passes the test.

4. Increase the variable of number of steps by one, if $j < b$ and $z \neq p - 1$, set $z \leftarrow z^2$.

5. If $z = 1$, then p is not a prime. If $z = p - 1$, then p is a probable prime and the test is passed. Otherwise, go back to step 4.

6. If $j = b$ and $z \neq p - 1$, then p is not a prime.

This method is faster than the previous one. This is because by Theorem 10.17, the probability that p is a probable prime is $\dfrac{3}{4}$.

10.4 THE RSA PROBLEM AND THE STRONG RSA PROBLEM

The RSA problem was proposed by Rivest, Shamir, and Adleman in 1978 when they published their famous RSA public key algorithm [7]. To a certain extent, it reflects the level of security of the RSA encryption algorithm (and the digital signature algorithm). Though no one is able to prove whether the hardness of solving the RSA problem is the same as that for large integer factorization, it is believed that the RSA problem is a very challenging problem. The RSA problem is thus used as the base of designing secure cryptographic algorithms. In other words, if the hardness of breaking RSA algorithms is equivalent to that of solving the RSA problem, then the cryptographic algorithm is regarded as secure.

Let p, q be two large primes with similar lengths in binary representation. Note that $n = pq$ is called a RIPE composite number if n is of at least 1024 bits and both $p - 1$, $q - 1$ have large prime factors.

Definition 10.18 Let $n = pq$ be a RIPE composite and e a positive odd number such that $(e, \varphi(n)) = 1$. Given a random integer $c \in \mathbb{Z}_n^*$, the problem of finding integers m such that $m^e \equiv c \pmod{n}$ is called an RSA problem. □

From the definition, the RSA problem is the problem of finding eth root in $\mathbb{Z}_n^*$. The case is special here in that $n = pq$ and e is a positive odd integer. The RSA assumption states that there is no polynomial time algorithm to solve RSA problems. The corresponding large integer factorization assumption is that for n, there is no polynomial time algorithm to obtain the factorization $n = pq$.

Based on the RSA problem, Baric and Pfitzmann [9], Fujisaki and Okamoto [10] defined the Strong RSA problem in 1997 independently.

Definition 10.19 Let $n = pq$ be the modulus of an RSA problem, G is a cyclic subgroup of $\mathbb{Z}_n^*$. Given a random $z \in G$, the problem of finding $(u, e) \in G \times \mathbb{Z}_n$ such that $z \equiv u^e \pmod{n}$ is called a Strong RSA problem. □

The Strong RSA problem is actually the problem of finding an arbitrary power modulo n. The Strong RSA hypothesis states that for any polynomial $P(l)$, the probability of finding solution $(u, e) \in G \times \mathbb{Z}_n$ to $z \equiv u^e \pmod{n}$ in polynomial time is less than $\frac{1}{P(l)}$, where l is the binary length of n.

In general, cryptographic algorithms based on RSA problem and derived RSA problem (such as Strong RSA problem) are usually classified into the category of integer factorization based algorithms. With the definition of RSA class problems, integer factorization–based algorithms can have a more flexible design, more variety of features, and the security of cryptographic algorithms. For example, many cryptographic algorithms with practical applications such as blind signatures, group signatures, and digital cash are based on RSA problems and derived RSA problems.

10.5 QUADRATIC RESIDUES

We have discussed the problems of quadratic residues in great detail previously. Since quadratic residues have many applications in cryptography, we shall further discuss special properties of quadratic residues that are involved in cryptography. Unless otherwise specified, the quadratic residues described in this section are on $\mathbb{Z}_n^*$, for the following n: $n = pq$ and $p \equiv 3 \pmod{4}, q \equiv 3 \pmod{4}$. The integer n with the above conditions is called a Blum number. We have the following result:

Theorem 10.20 *If n is a Blum number, then -1 is a quadratic nonresidue in $\mathbb{Z}_n^*$ and $\left(\dfrac{-1}{n}\right) = 1$.*

The proof is quite straightforward and is left to the reader.

This theorem is very important for some cryptographic algorithms and that is why the Blum number is introduced.

Definition 10.21 (*Quadratic Residue Assumption*) Let $n = pq$ be a product of two large primes. For a randomly selected $a \in QR_n$, finding x such that $x^2 \equiv a \pmod{n}$ is a hard problem. □

In this section, we prove that the quadratic residue assumption is probabilistically a polynomial equivalent to the factorization of n. For the quadratic residue problem modulo $n = pq$, we have a further result.

Theorem 10.22 *For any $a \in QR_n$, there are four square roots of a modulo n.*

Proof By the Chinese remainder theorem, solving $x^2 \equiv a \pmod{n}$ is equivalent to solving

$$\begin{cases} x^2 \equiv a \pmod{p}, \\ x^2 \equiv a \pmod{q}. \end{cases}$$

Let the roots of $x^2 \equiv a \pmod{p}$ and $x^2 \equiv a \pmod{q}$ be $\pm x_0$ and $\pm x_1$, respectively. Then the four roots of $x^2 \equiv a \pmod{n}$ are

$$\begin{cases} x \equiv \pm x_0 \pmod{p}, \\ x \equiv \pm x_1 \pmod{q}. \end{cases}$$

The theorem is proved.

Theorem 10.23 *For every quadratic residue, only one of the four square roots is a quadratic residue modulo n.*

Proof Let a be a quadratic residue modulo n. From the proof of Theorem 10.22, the four square roots of a can be assumed to be $\pm x, \pm y$, and

$$x \equiv y \pmod{p}, \quad x \equiv -y \pmod{q}.$$

Thus, the Jacobi symbols satisfy $\left(\dfrac{x}{n}\right) = -\left(\dfrac{y}{n}\right)$. Without loss of generality, we assume that x is the square root such that $\left(\dfrac{x}{n}\right) = 1$, then either $\left(\dfrac{x}{p}\right) = \left(\dfrac{x}{q}\right) = 1$ or $\left(\dfrac{-x}{p}\right) = \left(\dfrac{-x}{q}\right) = 1$. Thus, one of $x, -x$ must be a quadratic residue modulo n.

In summary, every quadratic residue a has four square roots x_1, x_2, x_3, x_4 which satisfy

$$\left(\frac{x_1}{p}\right) = \left(\frac{x_1}{q}\right) = 1, \quad \left(\frac{x_2}{p}\right) = \left(\frac{x_2}{q}\right) = -1,$$

$$\left(\frac{x_3}{p}\right) = -\left(\frac{x_3}{q}\right) = 1, \quad \left(\frac{x_4}{p}\right) = -\left(\frac{x_4}{q}\right) = -1.$$

The theorem is proved.

The importance of quadratic residues in cryptography is also reflected by the next result.

Theorem 10.24 *The problem of the factorization of n is probabilistically polynomial equivalent to solving the quadratic residue problem for n.*

Proof First, we need to show that if there is a polynomial time algorithm A such that for every input n, it outputs a factor of n, then there must be a polynomial time algorithm B such that for every input (n, x) with x a quadratic residue for n, it outputs a square root of x. The algorithm B is as follows:

1. Let $A(n) = p$, then $q = \dfrac{n}{p}$.

2. From Theorem 10.22, we can find four square roots of n.

Next, we need to show that if there is a polynomial time algorithm B such that for every input (n, x) with x a quadratic residue for n, it outputs a square root of x, then there must be a polynomial time algorithm A such that for every input n, it outputs a factor of n with probability $\frac{1}{2}$. The algorithm A is as follows:

1. Randomly select a number a such that $\left(\dfrac{a}{n}\right) = -1$. For an input $x \equiv y^2 \pmod{n}$, B outputs a square root b of x;

2. If $\left(\dfrac{b}{n}\right) = 1$, then a prime factor of n is obtained by computing $\gcd(a - b, n)$ or $\gcd(a + b, n)$.

The theorem is proved.

Finally, we state an assumption.

Definition 10.25 (*Decision problem of quadratic residuosity*) Let $n = pq$ be a product of two large primes. For a randomly selected $a \in \mathbb{Z}_n^*$, determining whether $a \in QR_n$ is a challenging problem. □

There has not been a proof of the hardness of the decision problem of quadratic residuosity modulo n. However, it is widely believed to be a difficult problem and is used in probabilistic encryption algorithms.

10.6 THE DISCRETE LOGARITHM PROBLEM

Another hard problem that is well used in cryptography is the discrete logarithm problem. The discrete logarithm problem and the integer factorization problem are the major hard problems from mathematics that support public key cryptography. Therefore, the public key cryptography constructions are divided into two classes according to these two kinds of problems—cryptography constructions based on the discrete logarithm and cryptography constructions based on the integer factorization. Like the case that the RSA problem derives the Strong RSA problem, the discrete logarithm problem also derives some hard problems and assumptions.

Definition 10.26 Let g be a primitive root in $\mathbb{Z}_p^*$ (which can also be stated as a generator of the cyclic group $\mathbb{Z}_p^*$). For any element $y \in \mathbb{Z}_p^*$, there is a unique x, $1 \leq x < p-1$ such that $g^x \equiv y \pmod{p}$. We call x the discrete logarithm modulo p of y with respect to the base g. □

In this case, the problem of finding a discrete logarithm is just the problem of finding an index. From the earlier discussion, we know that given $1 \leq x < p-1$, finding $y \in \mathbb{Z}_p^*$ such that $g^x \equiv y \pmod{p}$ is not difficult. However, so far there is no polynomial time algorithm for its inverse problem—the discrete logarithm problem. The best algorithm known to date is the NFS (the number field sieve method); its asymptotic time estimation is

$$e^{(1.923+O(1))(\ln(p))^{1/3}(\ln(\ln(p)))^{2/3}}.$$

By comparing the time estimations of the number field sieve method for both the discrete logarithm problem and the factorization of n, we see that they are of similar difficulty. At the present time, no one is able to tell which of these two problems is harder.

Definition 10.27 Let $g, h \in \mathbb{Z}_p^*$ be unrelated primitive roots of p (i.e., the discrete logarithm of g with respect to h is unknown). For any a with $(a,p) = 1$, we represent a as $a \equiv g^{\alpha}h^{\beta} \pmod{p}$, then (α, β) is called a representation of a modulo p with respect to the bases g, h. □

This definition can be extended to a case that involves more primitives.

Definition 10.28 Let $g_1, g_2, \cdots, g_s \in \mathbb{Z}_p^*$ be unrelated primitive roots of p. For any a with $(a,p) = 1$, we represent a as

$$a \equiv g_1^{\alpha_1} g_2^{\alpha_2} \cdots g_s^{\alpha_s} \pmod{p},$$

then $(\alpha_1, \alpha_2, \cdots, \alpha_s)$ is called a representation of a modulo p with respect to the bases $g_1, g_2, \cdots, g_s$. □

Such representations of a have important applications in group signatures and traceable blind signatures. The traceable blind signatures can be used to design protocols for digital cash.

Next, we will introduce an assumption that is related to a discrete logarithm problem.

Definition 10.29 (*Diffie–Hellman Problem*) Let g be a generator of the cyclic group $\mathbb{Z}_p^*$. For any $a, b \in \mathbb{Z}_p^*$ such that

$$a \equiv g^x \pmod{p}, \quad b \equiv g^y \pmod{p},$$

with unknown x, y, the Diffie–Hellman problem is to find c such that $c \equiv g^{xy} \pmod{p}$. □

Like the case in which the RSA problem can be used to assess the security of the RSA encryption algorithm, the Diffie–Hellman problem (DH problem for short) is used to assess the security of Diffie–Hellman key exchange protocol. The difficulty of the DH problem has not been determined at the moment. It is generally believed that the DH problem is hard. The DH problem and the discrete logarithm problem have the following relation: If the discrete logarithm problem can be solved in polynomial time, then the DH problem can also be computed in polynomial time. The converse is not necessarily true. However, it has been proved that under certain conditions, these two problems are equivalent. But even under these conditions, it is still difficult to solve the discrete logarithm problem; otherwise the discrete logarithm problem would be of no use in cryptography.

Finally, we describe the discrete logarithm problem for elliptic curves. As an example of finite groups, we have defined elliptic curves and introduced the groups of rational points on elliptic curves in Chapter 8. The structures and special properties of this kind of group, especially the discrete logarithm problem for the groups, is of particular importance in cryptography. These are theoretical bases for designing cryptographic algorithms using elliptic curves. Elliptic curve cryptographic algorithms are a popular topic of cryptography research. Some of the elliptic curves used in cryptography are of the following form.

Definition 10.30 Let $p > 3$ be a prime. The elliptic curve $E_p(a, b)$ defined over the finite field $\mathbb{Z}_p^*$ is given by the solutions $(x, y) \in \mathbb{Z}_p^* \times \mathbb{Z}_p^*$ of the equation

$$y^2 = x^3 + ax + b,$$

together with the point O of infinity. □

Definition 10.31 Let G be a cyclic subgroup of the elliptic curve $E_p(a, b)$ over the finite field $\mathbb{F}_p^*$. Let P be a generator of the group G, that is, $G = (P)$. The elliptic curve discrete logarithm problem is: given any point $Q \in G$, find a positive integer l, such that $Q = lP$. □

CHAPTER 11

Basics of Lattices

11.1 BASIC CONCEPTS

Let $\mathbb{R}$ be the real field and $\mathbb{R}^m$ the Euclidean space of dimension m. We use column vectors to represent the elements of $\mathbb{R}^m$ and define the inner product on $\mathbb{R}^m$ as

$$\langle . \rangle \quad \mathbb{R}^m \times \mathbb{R}^m \longmapsto \mathbb{R}$$
$$(\mathbf{x}, \mathbf{y}) \longmapsto \mathbf{x}^T \mathbf{y}.$$

The length $||\cdot||$ of vectors in $\mathbb{R}^m$ is derived from this inner product, that is, $\forall \mathbf{x} \in \mathbb{R}^m$, $||\mathbf{x}|| = \sqrt{\mathbf{x}^T \mathbf{x}}$.

Let $\mathbf{b}_1, \mathbf{b}_2, \cdots, \mathbf{b}_n$ be n linearly independent vectors of $\mathbb{R}^m$ $(m \geq n)$. The set

$$\mathcal{L}(\mathbf{b}_1, \mathbf{b}_2, \cdots, \mathbf{b}_n) = \left\{ \sum_{i=1}^{n} x_i \mathbf{b}_i \,|\, x_i \in \mathbb{Z} \right\}$$

is called a lattice of $\mathbb{R}^m$ and is simply denoted as L. The set $\{\mathbf{b}_1, \mathbf{b}_2, \cdots, \mathbf{b}_n\}$ is called a basis of L, m is called the dimension of L, and n is called the rank of L. A basis of the lattice L can also be written as a matrix form, which is the matrix $\mathbf{B} = [\mathbf{b}_1, \mathbf{b}_2, \cdots, \mathbf{b}_n] \in \mathbb{R}^{m \times n}$ whose column vectors are $\mathbf{b}_1, \mathbf{b}_2, \cdots, \mathbf{b}_n$. In this case, the lattice L is

$$\mathcal{L}(\mathbf{B}) = \{\mathbf{B}\mathbf{x} \,|\, \mathbf{x} \in \mathbb{Z}^n\}.$$

The determinant of L is defined to be $\det(L) = \sqrt{\det \mathbf{B}^T \mathbf{B}}$. The lattice L is called n dimensional full rank if $m = n$. In this case, the determinant of L is the absolute value of the determinant of the matrix $\mathbf{B}$, that is, $\det(L) = |\det(\mathbf{B})|$. In fact, $\det(L)$ is independent of the choice of basis of the lattice.

For linearly independent vectors $\mathbf{b}_1, \mathbf{b}_2, \cdots, \mathbf{b}_n$ of $\mathbb{R}^m$, we can perform orthogonalization process to get orthogonal vectors $\mathbf{b}_1^*, \mathbf{b}_2^*, \cdots, \mathbf{b}_n^*$. The Gram-Schmidt process can be summarized as follows: Let $\{\mathbf{b}_1, \mathbf{b}_2, \cdots, \mathbf{b}_n\}$ be a basis of a lattice L of $\mathbb{R}^m$. Define $\mathbf{b}_1^* = \mathbf{b}_1$,

$$\mathbf{b}_i^* = \mathbf{b}_i - \sum_{j=1}^{i-1} \mu_{i,j} \mathbf{b}_j^*, \ i > 1,$$

where

$$\mu_{i,j} = \frac{<\mathbf{b}_i, \mathbf{b}_j^*>}{<\mathbf{b}_j^*, \mathbf{b}_j^*>}, \ 1 \le j < i \le n.$$

Then $\{\mathbf{b}_1^*, \mathbf{b}_2^*, \cdots, \mathbf{b}_n^*\}$ is a set of orthogonal vectors of $\mathbb{R}^m$. It is noted that $\{\mathbf{b}_1^*, \mathbf{b}_2^*, \cdots, \mathbf{b}_n^*\}$ is usually not a basis of the lattice L. It is also noted that the determinant of L satisfies

$$\det(L) = \prod_{1 \le i \le n} ||\mathbf{b}_i^*||, \ \det(L) \le \prod_{1 \le i \le n} ||\mathbf{b}_i||.$$

11.2   SHORTEST VECTOR PROBLEM

The shortest vector problem (SVP) in lattices is the problem of finding a shortest nonzero vector of a lattice. This problem has been proved to be NP-hard for randomized reductions. In this section, we describe the existence theorem of the shortest vectors, that is, Minkowski's first theorem, and present an upper bound of a shortest vector.

First, we consider the existence of nonzero lattice vectors in a special region by stating the following theorem of Minkowski.

Theorem 11.1 *Let L be a lattice of $\mathbb{R}^m$ and S a measurable convex subset (i.e., $\alpha, \beta \in S \Rightarrow \frac{1}{2}(\alpha + \beta) \in S$) of $\mathbb{R}^m$ which is symmetric with respect to the origin (i.e., $a \in S \Rightarrow -a \in S$). If the volume of S satisfies $\mu(S) \ge 2^n \det(L)$, then $S \cap L$ contains a nonzero vector.*

We refer the reader to [11] for a proof.
We can use this theorem to prove Minkowski's first theorem.

Theorem 11.2 *Let L be a lattice of $\mathbb{R}^m$ with rank n and the length of the shortest lattice vector of L be λ_1. Then $\lambda_1 < \sqrt{n} \det(L)^{\frac{1}{n}}$.*

Proof Let $\{\mathbf{b}_1, \mathbf{b}_2, \cdots, \mathbf{b}_n\}$ be a basis of the lattice L and set $\mathbf{B} = [\mathbf{b}_1, \mathbf{b}_2, \cdots, \mathbf{b}_n] \in \mathbb{R}^{m \times n}$. The linear space generated by $\mathbf{b}_1, \mathbf{b}_2, \cdots, \mathbf{b}_n$ is denoted by $\mathrm{span}(\mathbf{b}_1, \mathbf{b}_2, \cdots, \mathbf{b}_n)$, that is, $\mathrm{span}(\mathbf{b}_1, \mathbf{b}_2, \cdots, \mathbf{b}_n) = \{\mathbf{B}\mathbf{x} | \mathbf{x} \in \mathbb{R}^n\}$.

Let $S = \mathcal{B}(O, \sqrt{n} \det(L)^{\frac{1}{n}}) \bigcap \mathrm{span}(\mathbf{b}_1, \mathbf{b}_2, \cdots, \mathbf{b}_n)$ be the open ball in $\mathrm{span}(\mathbf{b}_1, \mathbf{b}_2, \cdots, \mathbf{b}_n)$ with its center at the origin and with radius $\sqrt{n} \det(L)^{\frac{1}{n}}$. Note that the volume of S is strictly greater than $2^n \det(L)$ because S contains an n-dimensional hypercube with side length $2 \det(L)^{\frac{1}{n}}$. So by Theorem 11.1, there exists a nonzero vector $\mathbf{v} \in S \cap L$, such that $||\mathbf{v}|| < \sqrt{n} \det(L)^{\frac{1}{n}}$.

Therefore, the length λ_1 of a shortest lattice vector of L satisfies

$$\lambda_1 \leq ||\mathbf{v}|| < \sqrt{n} \det(L)^{\frac{1}{n}},$$

and the theorem is proved.

This theorem gives an upper bound of the length of shortest lattice vectors. Even though it is only a theorem of existence, the upper bound has significant practical interest.

11.3   LATTICE BASIS REDUCTION ALGORITHM

Minkowski's first theorem from Section 11.2 established an upper bound of the shortest lattice vectors, but no concrete construction of shorter vectors was given. In this section, we introduce the concept of reduced basis and describe the lattice basis reduction algorithm, which will be used to find shorter vectors in a lattice. In 1982, Lenstra et al. [12] designed a lattice basis reduction algorithm by which a reduced basis of a lattice L can be constructed. This algorithm is called the LLL algorithm. For convenience, we will assume the lattice L discussed in this section to be of n dimensional and full rank.

Definition 11.3 Define the map π_i from $\mathbb{R}^n$ to $\mathrm{span}(\mathbf{b}_1^*, \mathbf{b}_2^*, \cdots, \mathbf{b}_n^*)$ as

$$\pi_i(\mathbf{x}) = \sum_{j=i}^{n} \frac{<\mathbf{x}, \mathbf{b}_j^*>}{<\mathbf{b}_j^*, \mathbf{b}_j^*>} \mathbf{b}_j^*. \qquad \square$$

In fact, for any $\mathbf{x} \in \mathrm{span}(\mathbf{b}_1, \mathbf{b}_2, \cdots, \mathbf{b}_n)$, $\pi_i(\mathbf{x})$ is a component of the vector $\mathbf{x}$ that is orthogonal to each of the vectors $\mathbf{b}_1, \mathbf{b}_2, \cdots, \mathbf{b}_{i-1}$. In particular, $\pi_i(\mathbf{b}_i) = \mathbf{b}_i^*$.

Definition 11.4 A basis $\{\mathbf{b}_1, \mathbf{b}_2, \cdots, \mathbf{b}_n\}$ is said to be LLL-reduced if

1. $|\mu_{i,j}| \le \frac{1}{2}, 1 \le j < i \le n$.

2. $\frac{3}{4}||\pi_i(\mathbf{b}_i)||^2 \le ||\pi_i(\mathbf{b}_{i+1})||^2, 1 \le i < n$. ☐

Since $\pi_i(\mathbf{b}_i) = \mathbf{b}_i^*$, $\pi_i(\mathbf{b}_{i+1}) = \mathbf{b}_{i+1}^* + \mu_{i+1,i}\mathbf{b}_i^*$, the second requirement of the above definition can also be written as

$$\frac{3}{4}||\mathbf{b}_i^*||^2 \le ||\mathbf{b}_{i+1}^* + \mu_{i+1,i}\mathbf{b}_i^*)||^2, 1 \le i < n.$$

An LLL-reduced basis has the following properties:

Theorem 11.5 *If* $\{\mathbf{b}_1, \mathbf{b}_2, \cdots, \mathbf{b}_n\}$ *is an LLL-reduced basis of an n-dimensional lattice L, then*

1. $\det(L) \le \prod_{i=1}^{n} ||\mathbf{b}_i|| \le 2^{\frac{n(n-1)}{4}} \det(L)$.

2. $||\mathbf{b}_j|| \le 2^{\frac{i-1}{2}} ||\mathbf{b}_i^*||, 1 \le j < i \le n$.

3. $||\mathbf{b}_1|| \le 2^{\frac{n-1}{4}} \det(L)^{\frac{1}{n}}$.

4. *For any set of linearly independent vectors* $\mathbf{x}_1, \mathbf{x}_2, \cdots, \mathbf{x}_t$ *of L, we must have*

$$||\mathbf{b}_j|| \le 2^{\frac{n-1}{2}} \max\{||\mathbf{x}_1||, ||\mathbf{x}_2||, \cdots, ||\mathbf{x}_t||\}, 1 \le j \le t.$$

In particular, for any vector $\mathbf{x}$ *of L,* $||\mathbf{b}_1|| \le 2^{\frac{n-1}{2}} ||\mathbf{x}||$ *holds.*

The proof is omitted (see [12] for detail).

Next, we describe the lattice basis reduction algorithm using a two-dimensional lattice as an example. In essence, this algorithm for two-dimensional lattices is the Gaussian algorithm (see [13]):

1. Reduction step: $\mathbf{b}_2 = \mathbf{b}_2 - c\mathbf{b}_1$, where $c = \left\lfloor \frac{\langle \mathbf{b}_2, \mathbf{b}_1 \rangle}{\langle \mathbf{b}_1, \mathbf{b}_1 \rangle} + \frac{1}{2} \right\rfloor$;

2. Exchange step: if $||\mathbf{b}_1|| > ||\mathbf{b}_2||$, swap $\mathbf{b}_1 \Leftrightarrow \mathbf{b}_2$.

3. If $(\mathbf{b}_1, \mathbf{b}_2)$ is not a reduced basis, repeat the above steps.

It is noted that after the reduction step, we have $|\mu_{2,1}| \le 1/2$. The idea used in the LLL algorithm that will be introduced below is similar, after reduction $|\mu_{i,j}| \le 1/2$ holds for all $i > j$. Run the reduction step and the exchange step alternately, and swap two adjacent vectors in the exchange step if necessary.

The LLL algorithm:

Input: A basis of the lattice L: $\mathbf{B} = [\mathbf{b}_1, \mathbf{b}_2, \cdots, \mathbf{b}_n] \in \mathbb{Z}^{n \times n}$.
Output: An LLL-reduced basis of L.
(Loop): for $i = 1, \cdots, n$, do
        for $j = i - 1, \cdots, 1$, do
            $\mathbf{b}_i \leftarrow \mathbf{b}_i - c_{i,j}\mathbf{b}_j$, where $c_{i,j} = \left\lfloor \frac{\langle \mathbf{b}_i, \mathbf{b}_j^* \rangle}{\langle \mathbf{b}_j^*, \mathbf{b}_j^* \rangle} + \frac{1}{2} \right\rfloor$
        if there exists an i such that $\frac{3}{4}||\pi_i(\mathbf{b}_i)||^2 > ||\pi_i(\mathbf{b}_{i+1})||^2$,
        then
            swap $\mathbf{b}_i$ and $\mathbf{b}_{i+1}$
            goto (Loop)
        else output $\mathbf{B}$.

Now we analyze the correctness of the LLL algorithm. We need to make sure that after the reduction step, the inequality $|\mu_{i,j}| \le 1/2$ holds for all $i > j$. We let $\mathbf{b}_1', \cdots, \mathbf{b}_n'$ be the set of temporary vectors in the iteration process, where $\mathbf{b}_1' = \mathbf{b}_1$. In the reduction step, each $\mathbf{b}_i'$ is obtained by subtracting a suitable integer multiple of $\mathbf{b}_j'$ from $\mathbf{b}_i$ ($j < i$). Since $\mathbf{B}'$ is obtained from $\mathbf{B} = [\mathbf{b}_1, \cdots, \mathbf{b}_n]$ through a series of elementary transforms, $\mathbf{B}$ and $\mathbf{B}'$ are equivalent. The exchange step simply produces a new order of the column vectors of $\mathbf{B}'$, so $\mathbf{B}'$ obtained after each iteration is still a basis of the input lattice L.

The orthogonal basis $\mathbf{B}^*$ generated by $\mathbf{B}$ will not change after the reduction step, that is, the transformation from $\mathbf{B}$ to $\mathbf{B}'$ will not change the orthogonal vectors $\mathbf{b}_i^*$. It can be proved that all Gram–Schmidt coefficients $\mu_{i,j}$ ($i > j$) for $\mathbf{B}'$ satisfy $|\mu_{i,j}| \le \frac{1}{2}$.

Therefore, after the reduction steps, the condition $|\mu_{i,j}| \le \frac{1}{2}$ holds. Next, we will prove that the second condition for the LLL-reduced basis holds. If for some i

$$\frac{3}{4}||\pi_i(\mathbf{b}_i)||^2 > ||\pi_i(\mathbf{b}_{i+1})||^2,$$

then we can swap $\mathbf{b}_i$ and $\mathbf{b}_{i+1}$.

After exchanging two vectors, there might be some $j < i$ such that $|\mu_{i,j}| > 1/2$. In this case, we repeat the reduction process.

If after the reduction step, no pair of consecutive vectors needs to be swapped, then the output **B** is an LLL-reduced basis. This **B** is equivalent to the input matrix, since in the process of obtaining the output matrix, we merely performed a series of elementary transforms with integer coefficients.

This means that the LLL algorithm terminates by producing an LLL-reduced basis. Let the input be $\mathbf{b}_1, \cdots, \mathbf{b}_n$ and $M = \max\{||\mathbf{b}_1||^2, \cdots, ||\mathbf{b}_n||^2\}$, then the time complexity of the LLL algorithms is $O(n^6 \log^3 M)$. See [12] for the proof.

By item 4 of Theorem 11.5, we have $||\mathbf{b}_1|| \leq 2^{\frac{n-1}{2}} \lambda_1$, where λ_1 is the length of a shortest vector of L. Thus, the length of the vector $\mathbf{b}_1$ of the LLL-reduced basis is at most $2^{\frac{n-1}{2}}$ times that of a shortest vector. The vector $\mathbf{b}_1$ is called a $2^{\frac{n-1}{2}}$-approximate shortest vector. In some practical problems, the approximate shortest vector produced by the LLL algorithm is sufficient for a solution; there is no need to actually search for the shortest lattice vector. Therefore, this algorithm has a wide range of applications. See the survey [14] for more detail. We remark that non-polynomial time algorithms are available for finding shorter vectors (or shortest vectors) in an n-dimensional lattice, see [15, 16]. Also, as we mentioned earlier, shortest vectors in two-dimensional lattices can be found by using the Gaussian algorithm.

11.4 APPLICATIONS OF LLL ALGORITHM

Coppersmith proposed a method for finding small solutions of modular polynomial equations by using the LLL algorithm. In this section, we describe how to find small solutions of modular polynomial equations in one variable.

Theorem 11.6 (*Coppersmith*) *Let N be an integer of unknown factorization and $b \geq N^\beta$ be a factor of N. Let $f_b(x)$ be a monic polynomial of degree δ in one variable. Then roots x_0 to the equation $f_b(x) = 0$ (mod b) that satisfy*

$$|x_0| \leq \frac{1}{2} N^{\frac{\beta^2}{\delta} - \varepsilon}$$

can be found in time $O(\delta^6 \varepsilon^{-7} \log^3 N)$, where $\varepsilon > 0$ is an arbitrarily small number.

The basic idea of Coppersmith's method is to turn a modular polynomial equation into a polynomial equation of integer coefficients: Using $f(x)$ to construct $g(x)$ (with integer coefficients) such that the set of roots of $g(x)$ contains all small modular roots of $f(x)$, that is,

$$f(x_0) = 0 \pmod{b} \Rightarrow g(x_0) = 0, |x_0| \leq X.$$

Coppersmith's method solves the modular equation in two steps:

1. For a fixed integer m, construct a set of polynomials $C = \{f_1(x), \cdots, f_n(x)\}$ such that $f_i(x) = 0 \pmod{b^m}(i = 1, \cdots, n)$ have common solution x_0. For example, we can construct C as follows:

$$
\begin{aligned}
f_i(x) &= N^{m-i} f^i(x), i = 1, \cdots, m, \\
f_{m+i}(x) &= x^i f^m(x), i = 1, \cdots, m.
\end{aligned}
\tag{11.1}
$$

2. Using a linear combination of $\{f_i(x)\}_{i=1}^{n}$ to form $g(x) = \sum_{i=1}^{n} a_i f_i(x)$, $a_i \in \mathbb{Z}$, and ask $g(x)$ to satisfy

$$|g(x_0)| < b^m.$$

Since b^m divides each $f_i(x_0)$ for $i = 1, \cdots, n$, so b^m divides $g(x_0)$, that is, $g(x_0) = 0 \pmod{b^m}$. Note that $|g(x_0)| < b^m$ so $g(x_0) = 0$ holds.

The following lemma of Howgrave-Graham can be used to turn the problem of solving a modular polynomial equation with small roots to that of solving a polynomial equation with integer roots. For a polynomial $h(x) = \sum_i a_i x^i$, we define its Euclidean norm as $\|h(x)\| = \sqrt{\sum_i a_i^2}$.

Lemma 11.7 (*Howgrave-Graham*) *Let $h(x) \in \mathbb{Z}[x]$ be a polynomial with ω monomials. Let m be a positive integer. If*

1. $h(x_0) = 0 \bmod b^m$, *where* $|x_0 \leq X$.

2. $\|h(xX)\| < b^m / \sqrt{\omega}$.

Then $h(x_0) = 0$ holds.

Proof We have

$$|g(x_0)| = \sum_i c_i x_0^i \le \sum_i |c_i x_0^i| \le \sum_i |c_i| X^i \le \sqrt{\omega} \|g(xX)\| < b^m.$$

Since $g(x_0)$ is a multiple of b^m, so $g(x_0) = 0$.

Using the powers of $f(x)$, we construct a series of polynomials $f_1(x), \cdots, f_n(x)$, such that each modular equation $f_i(x) = 0 \pmod{b^m}$ $(i = 1, \cdots n)$ has x_0 as a solution. For example, we can use (11.1). If $g(x)$ is set to be a linear combination of $f_1(x), \cdots, f_n(x)$, then we have

$$g(x_0) = \sum_{i=1}^n a_i f_i(x_0) = 0 \pmod{b^m}, \text{ with } a_i \in \mathbb{Z}.$$

Each linear combination of $f_1(x), \cdots, f_n(x)$ satisfies condition (1) of Lemma 11.7. We need to find a linear combination that satisfies condition (2) of Lemma 11.7. In other words, we need to find a linear combination $g(x)$ whose Euclid norm is less than $b^m/\sqrt{n}$. Let L be the lattice generated by the coefficients vectors of functions $f_i(xX)$ $(i = 1, \cdots, n)$. We will use the LLL algorithm to find a shorter vector and get the desired $g(x)$.

The LLL algorithm enables us to find a nonzero lattice vector $\mathbf{v}$ of L such that $\|\mathbf{v}\| < b^m/\sqrt{n}$. In fact, let $\mathbf{v} = \mathbf{b}_1$, where $\mathbf{b}_1$ is the first vector in the LLL-reduced basis. By Theorem 11.5, we have

$$\|\mathbf{v}\| \le 2^{\frac{n-1}{4}} \det(L)^{\frac{1}{n}}.$$

$\det(L)$ can be computed by using the coefficients of $f_i(xX)$. If the following condition

$$2^{\frac{n-1}{4}} \det(L)^{\frac{1}{n}} < \frac{N^{\beta m}}{\sqrt{n}} \tag{11.2}$$

is satisfied, then we get $\|\mathbf{v}\| < \frac{N^{\beta m}}{\sqrt{n}} \le \frac{b^m}{\sqrt{n}}$. If the multiple in (11.2) is neglected, we have approximately that

$$\det(L) < N^{\beta mn}.$$

If an n-dimensional lattice L satisfies the above condition and $\mathbf{B}$ is a basis of L, then on average, the contribution of a basis vector $\mathbf{v} \in \mathbf{B}$ to $\det(L)$ is smaller than $N^{\beta m}$. A vector satisfying this condition is called a "usable" vector. We need to use "usable" vectors to construct a basis for the lattice.

Now we prove Theorem 11.6.

Proof Define $X = \frac{1}{2} N^{\frac{\beta^2}{\delta} - \varepsilon}$. As mentioned before, Coppersmith's method has two steps.

(1) Let $m = \left\lceil \frac{\beta^2}{\delta \varepsilon} \right\rceil$ and C be the set of the following polynomials:

$$
\begin{array}{ccccc}
N^m, & xN^m, & x^2 N^m, & \cdots, & x^{\delta-1} N^m, \\
N^{m-1} f, & xN^{m-1} f, & x^2 N^{m-1} f, & \cdots, & x^{\delta-1} N^{m-1} f, \\
N^{m-2} f^2, & xN^{m-2} f^2, & x^2 N^{m-2} f^2, & \cdots, & x^{\delta-1} N^{m-2} f^2, \\
\vdots, & \vdots, & \vdots, & & \vdots, \\
N f^{m-1}, & xN f^{m-1}, & x^2 N f^{m-1}, & \cdots, & x^{\delta-1} N f^{m-1}.
\end{array}
$$

We also need polynomials

$$
f^m, x f^m, x^2 f^m, \cdots, x^{t-1} f^m,
$$

where t is a parameter related to m whose value is determined later.

Note that the kth polynomial of C is a polynomial of degree k, and this introduces a new monomial x^k. We can also write C in the following form:

$$
g_{i,j}(x) = x^j N^i f^{m-i}(x), i = 0, \cdots, m-1, j = 0, \cdots, \delta - 1
$$
$$
h_i(x) = x^i f^m(x), i = 0, \cdots, t-1.
$$

(2) Next, we construct the lattice L from the coefficients of $g_{i,j}(xX)$ and $h_i(xX)$.

We can list $g_{i,j}$ and h_i according the ascending order of their degrees. The column vectors of a basis $\mathbf{B}$ of L are the coefficients vectors $g_{i,j}(xX)$ and $h_i(xX)$. Let $n = \delta m + t$, then $\mathbf{B}$ is the following $n \times n$ lower triangular matrix

$$\begin{pmatrix}
N^m & & & & & & & & & & & \\
 & N^m X & & & & & & & & & & \\
 & & \ddots & & & & & & & & & \\
 & & & \ddots & \ddots & \ddots & & \ddots & & & & \\
- & - & \cdots & - & \cdots & NX^{\delta m-\delta} & & & & & & \\
- & - & \cdots & - & \cdots & - & NX^{\delta m-\delta+1} & & & & & \\
 & & \ddots & \ddots & \ddots & \ddots & & \ddots & \ddots & & & \\
- & - & \cdots & - & \cdots & - & - & \cdots & NX^{\delta m+1} & & & \\
- & - & \cdots & - & \cdots & - & - & - & - & X^{\delta m} & & \\
- & - & \cdots & - & \cdots & - & - & - & - & - & X^{\delta m+1} & \\
 & & \ddots & \ddots & & \ddots & & \ddots & \ddots & \ddots & \ddots & \ddots \\
 & \ddots & \ddots & \ddots & - & - & - & - & - & - & \cdots & X^{\delta m+t-1}
\end{pmatrix}$$

Since the basis $\mathbf{B}$ of L is lower triangular, $\det(L)$ is the product of the diagonal elements of $\mathbf{B}$:

$$\det(L) = N^{\frac{1}{2}\delta m(m+1)} X^{\frac{1}{2}n(n-1)}. \tag{11.3}$$

Next, we determine the optimal parameter t, or the optimal $n = \delta m + t$. As discussed before, the vectors whose contribution to $\det(L)$ are smaller than $N^{\beta m}$ are "usable." In this case, it requires that the diagonal component of the coefficients vector of $h_i(xX)$ be less than $N^{\beta m}$, that is,

$$X^{n-1} < N^{\beta m}. \tag{11.4}$$

Since $X^{n-1} < N^{(\frac{\beta^2}{\delta}-\varepsilon)(n-1)} < N^{\frac{\beta^2}{\delta}n}$, so if we take

$$n \leq \frac{\delta}{\beta}m, \tag{11.5}$$

then (11.4) holds. Since $m = \lceil \frac{\beta^2}{\delta\varepsilon} \rceil$, we know that $m \leq \frac{\beta^2}{\delta\varepsilon} + 1$. Thus, the dimension of the lattice is such that

$$n \leq \frac{\beta}{\varepsilon} + \frac{\delta}{\beta}.$$

As $7\beta^{-1} \leq \varepsilon^{-1}$, we see that $n = O(\varepsilon^{-1}\delta)$. We let n be the smallest value such that (11.5) holds, then $n > \frac{\delta}{\beta} - 1 \geq \frac{\beta}{\varepsilon} - 1 \geq 6$.

Now we prove that the LLL algorithm produces the desired short vector. By Lemma 11.7, we need to use the LLL algorithm to find a vector $\mathbf{v}$ such that $\|\mathbf{v}\| \leq \frac{b^m}{\sqrt{n}}$. By the property of reduced basis, it is sufficient to make

$$2^{\frac{n-1}{4}} \det(L)^{\frac{1}{n}} < \frac{b^m}{\sqrt{n}}$$

to hold true.

From (11.3) and $b \geq N^\beta$, it is sufficient to show

$$N^{\frac{\delta m(m+1)}{2n}} X^{\frac{n-1}{2}} \leq 2^{-\frac{n-1}{4}} n^{-\frac{1}{2}} N^{\beta m},$$

that is,

$$X \leq \frac{1}{2} N^{\frac{2\beta m}{n-1} - \frac{\delta m(m+1)}{n(n-1)}}.$$

But $X = \frac{1}{2} N^{\frac{\beta^2}{\delta} - \varepsilon}$, so it is enough to prove

$$\frac{2\beta m}{n-1} - \frac{\delta m^2 (1 + \frac{1}{m})}{n(n-1)} \geq \frac{\beta^2}{\delta} - \varepsilon.$$

Multiply both sides of the above by $\frac{n-1}{n}$ and use $n \leq \frac{\delta}{\beta} m$, we get

$$2\frac{\beta^2}{\delta} - \frac{\beta^2}{\delta}^2 \left(1 + \frac{1}{m}\right) \geq \frac{\beta^2}{\delta} - \varepsilon,$$

that is,

$$-\frac{\beta^2}{\delta} \cdot \frac{1}{m} \geq -\varepsilon.$$

The above relation holds if $m \geq \frac{\beta^2}{\delta \varepsilon}$, our choice of $m = \left\lceil \frac{\beta^2}{\delta \varepsilon} \right\rceil$ satisfies this condition. Therefore, the desired vector can be found by the LLL algorithm.

Treat this short vector as the coefficients vector of the polynomial $g(x)$, then $g(x)$ satisfies the two conditions of Lemma 11.7. Therefore, solving a modular polynomial for small solutions x_0 is turned into finding a solution x_0 of the polynomial equation $g(x) = 0$. The latter is an easier computation task.

It can be proved that an upper bound of the lengths of vectors in $\mathbf{B}$ is $2^{O(\varepsilon^{-1} \log N)}$. This implies that the time complexity of using the LLL algorithm to solve the above modular polynomial equation is $O(\delta^6 \varepsilon^{-7} \log^3 N)$. The theorem is proved.

Setting $\varepsilon = \frac{1}{\log N}$, A. May [14] has the following improvement of the above theorem of Coppersmith [17].

Theorem 11.8 *Let N be an integer of unknown factorization and $d \geq N^{\beta}$ be a factor of N. Let $f_b(x)$ be a monic polynomial of degree δ in one variable and c_N be a quantity bounded by $\log N$. Then we can find solutions x_0 to the equation $f_b(x) = 0 \pmod{b}$ that satisfy*

$$|x_0| \leq c_N N^{\frac{\beta^2}{\delta}}$$

in polynomial time in $(\log N, \delta)$.

By Theorem 11.6, it is easy to show that if $N = pq$ with p, q having the same binary length, then knowing $\frac{1}{4} \log_2 N$ high bits (or low bits) implies that the factorization of N can be performed in polynomial time. This result was initially proved by Coppersmith [17] by using the method for finding small roots for bivariate polynomials.

Coppersmith's method of finding small roots for modular polynomials has many applications in the analysis of RSA encryption algorithms. See, for example, [18–20], and the survey article [14].

References

1. C. Pan and C. Pan. *Elementary Number Theory*. Beijing, Peking University Press, 2003 (in Chinese).

2. P. Wu. *Modern Algebra*. Beijing, People Education Press, 1979 (in Chinese).

3. H. Zhang. *The Foundation of Modern Algebra*. Beijing, People Education Press, 1978 (in Chinese).

4. S. Liu. *The Foundation of Modern Algebra*. Beijing, Higher Education Press, 1999 (in Chinese).

5. C. Pan and C. Pan. *The Foundation of Analytic Number Theory*. Beijing, Science Press, 1991 (in Chinese).

6. C. Pan and C. Pan. *Elementary Proofs of the Prime Number Theorem*. Shanghai, Shanghai Science and Technology Press, 1988 (in Chinese).

7. R. Rivest, A. Shamir, and L. Adleman. A method for obtaining digital signatures and public-key cryptosystems. *Communications of the ACM*, Vol. 21(2), pp. 120–126, 1978.

8. M.J. Wiener. Cryptanalysis of short RSA secret exponents. *IEEE Transactions on Information Theory*, Vol. 36, pp. 553–558, 1990.

9. N. Baric and B. Pfitzmann. Collision-free accumulators and fail-stop signature schemes without trees. *Advances in Cryptology—EUROCRYPT 1997*, LNCS 1233, pp. 480–494, 1997.

10. E. Fujisaki and T. Okamoto. Statistical zero knowledge protocols to prove modular polynomial relations. *Advances in Cryptology—CRYPTO 1997*, LNCS 1294, pp. 16–30, 1997.

11. K. Feng. *Algebraic Number Theory*. Beijing, Science Press, 2000 (in Chinese).

12. A.K. Lenstra, H.W. Lenstra, and L. Lovász. Factoring polynomials with rational coefficients. *Mathematische Annalen*, Vol. 261, pp. 513–534, 1982.

13. D. Micciancio and S. Goldwasser. *Complexity of Lattice Problems: A Cryptographic Perspective*. Boston, Kluwer Academic Publishers, 2002.

14. A. May. Using LLL-reduction for solving RSA and factorization problems: A survey. In *The LLL Algorithm, Survey and Applications* (Editors: P. Nguyen and B. Vallee), Heidelberg, Springer, 2009, pp. 315–348.

15. M. Ajtai, R. Kumar, and D. Sivakumar. A sieve algorithm for the shortest lattice vector problem. In *Proceedings on 33rd Annual ACM Symposium on Theory of Computing*, STOC 2001, 266–275, Heraklion, Crete, Greece. ACM.

16. R. Kannan. Minkowski's convex body theorem and integer programming. *Mathematics of Operation Research*, Vol. 12(3), pp. 415–440, 1987.

17. D. Coppersmith. Small solutions to polynomial equations and low exponent RSA vulnerabilities. *Journal of Cryptology*, Vol. 10(4), pp. 223–260, 1997.

18. D. Boneh and G. Durfee. Cryptanalysis of RSA with private key d less than $N^{0.292}$. *IEEE Transactions on Information Theory*, Vol. 46(4), pp. 1339–1349, 2000.

19. J.S. Coron and A. May. Deterministic polynomial-time equivalence of computing the RSA secret key and factoring. *Journal of Cryptology*, Vol. 20(1), pp. 39–50, 2007.

20. E. Jochemsz and A. May. A strategy for finding roots of multivariate polynomials with new applications in attacking RSA variants. *Advances in Cryptology*–ASIACRYPT 2006, LNCS 4284, pp. 267–282, 2006.

Further Reading

L.K. Hua. *Introduction to Number Theory.* Beijing, Science Press, 1979 (in Chinese).

C. Min and S. Yan. *Elementary Number Theory.* Beijing, Higher Education Press, 2003 (in Chinese).

L. Nie and S. Ding. *Introduction to Algebra.* Beijing, Higher Education Press, 2000 (in Chinese).

X. Wang. *Lecture Notes in Number Theory and Algebraic Structures.* 2003.

Index

Milton Keynes UK
Ingram Content Group UK Ltd.
UKHW031131141024
449569UK00006B/278